T0211805

Heterotopia and Globalisation in the Twenty-First Century

Can heterotopia help us make sense of globalisation? Against simplistic visions that the world is becoming one, *Heterotopia and Globalisation in the Twenty-First Century* shows how contemporary globalising processes are riven by heterotopian tension and complexities.

A heterotopia, in Michel Foucault's initial formulations, describes the spatial articulation of a discursive order, manifesting its own distinct logics and categories in ways that refract or disturb prevailing paradigms. While in the twenty-first century the concept of globalisation is frequently seen as a tumultuous *un*differentiation of cultures and spaces, this volume breaks new ground by interrogating how heterotopia and globalisation in fact intersect in the cultural present. Bringing together contributors from disciplines including Geography, Literary Studies, Architecture, Sociology, Film Studies, and Philosophy, this volume sets out a new typology for heterotopian spaces in the globalising present. Together, the chapters argue that digital technologies, climate change, migration, and other globalising phenomena are giving rise to a heterotopian multiplicity of discursive spaces, which overlap and clash with one another in contemporary culture.

This volume will be of interest to scholars across disciplines who are engaged with questions of spatial difference, globalising processes, and the ways they are imagined and represented.

Simon Ferdinand runs English Academic Editing (eaediting.nl). Having received his PhD *cum laude* from the University of Amsterdam, he is the author of *Mapping Beyond Measure: Art, Cartography, and the Space of Global Modernity*, and co-editor of *Other Globes: Past and Peripheral Imaginations of Globalisation*.

Irina Souch is Lecturer in the Department of Modern Foreign Languages and Cultures of the University of Amsterdam and Research Fellow at the Amsterdam School of Cultural Analysis (ASCA). She is author of *Popular Tropes of Identity in Contemporary Russian Television and Film* (Bloomsbury 2017). Her current work addresses narrative, aesthetic, and political functions of landscape in serial television drama and film.

Daan Wesselman is Lecturer in Literary and Cultural Analysis at the University of Amsterdam. Always trying to bridge the humanities and urban studies, he focuses particularly on postindustrial urban redevelopment and the material-discursive interfaces between the body, the city, and everyday life.

Heterotopia and Globalisation in the Twenty-First Century

**Edited by Simon Ferdinand,
Irina Souch and Daan Wesselman**

Routledge
Taylor & Francis Group

LONDON AND NEW YORK

First published 2020
by Routledge
2 Park Square, Milton Park, Abingdon, Oxon OX14 4RN

and by Routledge
605 Third Avenue, New York, NY 10017

Routledge is an imprint of the Taylor & Francis Group, an informa business

First issued in paperback 2021

Publisher's Note
The publisher has gone to great lengths to ensure the quality of this reprint but points out that some imperfections in the original copies may be apparent.

Frontispiece.
Heterotopia, Henrietta Simson, 2011, 210 x 40cm, width variable, wood, canvas, gesso, pigment and A4 photocopy of *Castle by a Lake*, Ambrogio Lorenzetti, c.1340, Pinacoteca Nazionale, Siena, 32.5 x 22.5cm, tempera on panel
© Henrietta Simson

British Library Cataloguing-in-Publication Data
A catalogue record for this book is available from the British Library

Library of Congress Cataloging-in-Publication Data
A catalog record for this book has been requested

ISBN 13: 978-0-367-25956-3 (hbk)
ISBN 13: 978-0-429-29073-2 (ebk)
ISBN 13: 978-1-03-223865-4 (pbk)

Typeset in Times New Roman
by Apex CoVantage, LLC

Contents

Editors and contributors

Editors

Simon Ferdinand is an interdisciplinary researcher whose interests sit between visual culture, geography, and globalisation studies. His monograph, *Mapping Beyond Measure: Art, Cartography and the Space of Global Modernity*, appeared with the University of Nebraska Press in 2019. He is the author of many articles and chapters on artistic mapping practices and cultural visions of globalisation and the global, and co-editor of *Other Globes: Past and Peripheral Imaginations of Globalisation*. He runs a developmental editing company (www.eaediting.nl).

Irina Souch is Lecturer in the Department of Modern Foreign Languages and Cultures of the University of Amsterdam and Research Fellow at the Amsterdam School of Cultural Analysis (ASCA). She is author of *Popular Tropes of Identity in Contemporary Russian Television and Film* (Bloomsbury 2017). Her current work addresses narrative, aesthetic, and political functions of landscape in serial television drama and film which requires exploration of various aspects of contemporary critical theory, philosophy, cultural geography, and film and television studies.

Daan Wesselman is Lecturer in Literary and Cultural Analysis at the University of Amsterdam. As a researcher affiliated with the Amsterdam School for Cultural Analysis, his focus is on finding common ground between the humanities and urban studies. He writes about concepts such as heterotopia, nonplace, and the posthuman as a means of understanding postindustrial urban redevelopment and the material-discursive interfaces between the body, the city, and everyday life. His work appears in *Space and Culture* and several edited books.

Contributors

Elham Bahmanteymouri is Lecturer in Urban Planning at the University of Auckland, New Zealand, where she completed her PhD in planning. She specialises in urban economics, urban critical theory, and behavioural economics. Through deployment of a Lacanian (post)Marxist approach, her research is concerned with understanding urban phenomena and suggesting better solutions to urban problems such as provision of affordable housing and implementation and evaluation of urban plans and policies. Elham also has extensive working experience as a senior planner in both public and private sectors in Iran for sixteen years.

Adam Kildare Cottrel is Assistant Professor of Film at Georgia Gwinnett College. He works at the intersection of film and philosophy. He is especially interested in the aesthetic dimension of film, particularly how style can give form to logics of power through its mediation of cinematic space, time, and bodies. In addition, current research projects include a theory of endurance and the politics of survival, a study on the form and philosophy of liquidity, as well as an inquiry into the relation between film style and globalisation. His work has appeared in *Paragraph*, *World Picture*, and *liquid blackness*, amongst others.

Lieven De Cauter is a philosopher, art historian, and activist who teaches in the Department of Architecture, Urbanism and Planning at KU Leuven and at RITCS in Brussels. He has published extensively on experience and modernity, architecture and politics. He is Co-editor, with Michiel Dehaene, of the volume *Heterotopia and the City: Public Space in a Postcivil Society* (Routledge, 2008). His other books include *Art and Activism in the Age of Globalization* (co-edited with Ruben de Roo and Karel Vanhaesebrouck, nai010, 2011), and *Entropic Empire: On the City in the Age of Disaster* (nai010, 2012).

Cathy Elliott is Senior Teaching Fellow at UCL. She is interested in ideas about time, temporality, and story-telling in politics and international relations. Her first book, *Democracy as Foreign Policy: Temporal Othering in International Relations*, came out in 2017; she has also published articles and books chapters using Foucault's thought to critique various stories we tell about democracy. She is recapitulating some of those themes, and finding new ones, in her current book project on the British "new nature writing." She is interested in the politics of pedagogy and also writes about teaching and learning.

Mary Gearey is Senior Research Fellow within the University of Brighton's School of Environment and Technology. Her research seeks to bridge the disjuncture between water resources management practice and local environmentalism. Her current work is critically engaged with understanding how developed economies organise and manage their freshwater resources with regards to transitioning towards sustainable futures in the context of climate change. Her most recent work appeared in *Geography*, *Sustainable Development* and several edited volumes.

Farzaneh Haghighi is Lecturer in Architecture at the School of Architecture and Planning, University of Auckland, New Zealand. She holds a PhD in Architecture from The University of Sydney, Australia. Her research is concerned with the intersection of political philosophy, architecture, and urbanism, and her first book, *Is the Tehran Bazaar Dead? Foucault, Politics, and Architecture*, was published in July 2018 by Cambridge Scholars Publishing. Her research seeks new avenues to enrich our creative analysis of complex built environments through investigating the implications of critical and cultural theory for architectural knowledge.

Kevin Hetherington is Professor of Geography and Pro-Vice-Chancellor Research, Enterprise and Scholarship at the Open University. He has published extensively on museums, heritage, consumption, urban regeneration, and spatial theory. His book publications include *The Badlands of Modernity: Heterotopia and Social Ordering* (Routledge, 1997), *Capitalism's Eye: Cultural Spaces of the Commodity* (Routledge, 2007), *Consuming the Entrepreneurial City: Image, Memory, Spectacle* (co-edited with Anne Cronin; Routledge, 2008), and *Urban Rhythms: Mobilities, Space and Interaction in the Contemporary City* (co-edited with Robin Smith; Wiley, 2013).

Peter Johnson completed his interdisciplinary PhD "On Heterotopia" at the University of Bristol. He has published a variety of journal articles related to Foucault's use of space as a tool of analysis and a range of specific heterotopian sites, including cemeteries, gardens, and ships. In 2012, he set up the website "Heterotopian Studies" which has become internationally recognised as a site for academics, writers, and artists, sharing and developing resources and ideas relating to Foucault's ideas on heterotopia. His most recent independent research involves exploring how the digital domain impacts on formulations of heterotopia.

Ursula Kluwick is Senior Assistant of Modern English Literature at the University of Bern. She holds a doctoral degree from the University of Vienna and completed her habilitation on water in the Victorian age at the University of Bern in 2017. From 2014–2016, she was a Marie Heim-Vögtlin fellow of the Swiss National Science Foundation. Her research focuses on ecocriticism and postcolonial studies, and her publications include *Exploring Magic Realism in Salman Rushdie's Fiction* (Routledge 2011), and *The Beach in Anglophone Literatures and Cultures: Reading Littoral Space*, edited with Virginia Richter (Ashgate 2015). She is currently editing an interdisciplinary handbook on sustainability with Evi Zemanek (UTB 2019) and preparing her monograph on *Ecological Hauntings* for publication. Her current research focuses on the Mediterranean beach as fugitive space.

Gladys Pierpauli received her MA degree in Cultural Sociology form the National University of San Martin in Argentina. She is currently Professor at ISER (Instituto Su- perior de EnsenÞanza Radiofoìnica, Argentina). Her latest journal article, "Transnational Hybridity: Argentine film representation of Chinese(ness)," was published in *Glocalism*.

Virginia Richter is Full Professor of Modern English Literature at the University of Bern. She holds a doctoral degree in Comparative Literature from the University of Munich, where she also completed her habilitation on literary representations of Darwinism. She was a Visiting Fellow at the University of

Kent at Canterbury and at the University of Leeds, and a Visiting Professor at the University of Göttingen. Her most recent publications include *The Beach in Anglophone Literatures and Cultures: Reading Littoral Space*, edited with Ursula Kluwick (Ashgate 2015), and a special issue of the *European Journal of English Studies* (*EJES*) on "Modern Creatures," edited with Pieter Vermeulen (2015). In spring 2018, she was a Visiting Research Fellow at the IASH, University of Edinburgh, where she worked on a monograph about the beach in modernist literature. She is also preparing a research project on the beach in the long twentieth century.

Henrietta Simson is an artist whose work explores the landscape image through its historical and cultural development, and its current definition within a digital context framed by ecological crisis. She completed a practice-related PhD in 2017, thesis title: *Landscape After Landscape, Pre-Genre Backgrounds in a Post-Genre Digital Age*. She received the Clare Winsten Memorial Award and the Gordon Luton Award for Fine Art in 2007, and in 2011 was awarded the Threadneedle Prize for Painting and Sculpture. Her recent research has been published by I.B. Tauris, Cambridge Scholars Publishing, *Aesthetica Magazine*, University of Minnesota Press, and the Courtauld Institute, London. She is currently teaching at the University of the Arts, London.

Hanneke Stuit is Assistant Professor of Literary and Cultural Analysis at the University of Amsterdam and Researcher at the Amsterdam School of Cultural Analysis (ASCA). She works on peripheral spaces in the globalised present, particularly on the imaginaries generated about carceral and rural spaces. She focuses on the crucial role played by cultural imaginations in determining what aspects of contemporary carceral and rural life do and do not become visible nationally and globally and on how these imaginations can be mobilised politically. Her main areas of interest are South Africa and the Netherlands.

Graham St John, PhD, is a cultural anthropologist and currently Senior Research Fellow on a comparative ethnography of Burning Man Regional Events, supported by the SNSF and based in Social Science at the University of Fribourg, Switzerland. Among Graham's eight books are *Mystery School in Hyperspace: A Cultural History of DMT* (North Atlantic Books 2015), *Global Tribe: Technology, Spirituality and Psytrance* (Equinox 2012), *Technomad: Global Raving Countercultures* (Equinox, 2009), and the edited collections *Weekend Societies: Electronic Dance Music Festivals and Event-Cultures* (Bloomsbury 2017) and *Victor Turner and Contemporary Cultural Performance* (Berghahn 2008). He is Executive Editor of *Dancecult: Journal of Electronic Dance Music Culture*.

Mariano Turzi received his PhD from the School of Advanced International Studies at Johns Hopkins University and is currently a Professor at the University of CEMA in Argentina. He has written on the international political economy of natural resources. His latest book is *The Political Economy of Agricultural Booms: Managing Soybean Production in Argentina, Brazil, and Paraguay* (Palgrave, 2017).

Acknowledgments

First and foremost, we would like to thank the contributors for their ideas, patience, and cooperation, without which this book would not have been possible. We wholeheartedly thank Faye Leerink at Routledge for her editorial support, which proved invaluable. Thanks also to the anonymous reviewers for their encouraging comments and stimulating suggestions, which pushed us to develop crucial aspects of the book. In addition, we are grateful to the Amsterdam School for Cultural Analysis's Peripheries reading group for promoting this project and many inspiring discussions that fed into the points and perspectives explored in what follows. Lastly, our special thanks to Henrietta Simson for providing the image for the frontispiece which so wonderfully represents the spirit of this volume.

1 Introduction

Interrupting globalisation – heterotopia in the twenty-first century

Daan Wesselman, Simon Ferdinand, and Irina Souch

The idea for this book arose on a spring day in Amsterdam, on which we three editors met in a new coffee place just off Mercatorplein square. We sat among an eclectic mix of vintage chairs and wooden benches, sipping espressos made from beans roasted in the shop itself, according to their distinct "roasting philosophy." The shop had been redesigned to create intimate corners and in the centre of the space was a communal table for reading and working, a mass of untreated lumber. This gave the place a natural atmosphere – there were plenty of plants too – as well as a somewhat DIY or "pop-up store" feel. The clientele was relatively young and fashionable, and there was a high density of MacBooks in view. The male staff sported moustaches and plain clothing; the female staff septum piercings, bridging their nostrils. The coffee shop was new and hip, and everything about it was "now": from the facial-hair-du-jour look, through the vintage chairs and normcore clothing, to minimal-but-natural furnishings and interior design.

Of course, everyone – or at least everyone living in privileged global metropoles and (over)developed regions – knows a place like this: it is a textbook example of a gentrifiers' coffee shop, almost to the point of caricature. It ticks all the boxes for the global development of gentrification of poor neighbourhoods, from the product – an artisanal alternative to the more obviously globalising force that is Starbucks – through the eco-minimal aesthetics to the MacBooks for the globally mobile (if endemically precarious) creative class. In this sense, nothing about this coffee shop is specific to its locality – it might as well have been in Barcelona or Cape Town, Beijing or Mexico City.

The appeal of this particular coffee shop therefore goes beyond the coffee and fashionable aesthetics themselves, since those can be had anywhere. Despite being entirely formulaic, this coffee place is especially notable on account of the *difference* between it and the surrounding neighbourhood, the Mercatorplein. This square is where the Turkish residents of the city go to celebrate if the national football squad wins a big match, for example. Much of the area is dominated by social housing, a high proportion of which is inhabited by low-income residents with migrant backgrounds. Until recently, the high street was characterised by Turkish grocery stores, fastfood restaurants, and offices for international money transfers. The area has undergone significant changes in recent years: much like the new vintage clothing store and organic food take-away opposite, our coffee shop represents a departure from the kind of store one might have previously expected to find here. Seen in the context of this specific locality, therefore, the overwhelming impression and "selling point" of this coffee shop is difference: *different* in its products, *different* in its aesthetics, and *different* in its clientele.

In demonstrating how globalising processes often give rise to disjunctions among sites and subjects, this mundane example of the gentrified coffee shop opens up our concerns in this volume. The shop is simultaneously markedly different to its surroundings and entirely generic in its coffee shop formula; it is rooted in the local while also staunchly global; it harkens back to the traditional coffee house where novel ideas are exchanged (cf. Hetherington 1997) and the shop and its clientele are also undeniably accomplices of neoliberal capital. The result of these tensions is a mixture of simultaneous uniformity and difference. Dropped into, or working their way through, pregiven social worlds that they will never wholly displace, globalising phenomena such as the "hipsterised" coffee shop exist in states of discrepancy, dis-location, or dis-placement. Our point is not simply that globalising discourses move geographically as part of globalisation's great jostling about of people, places, and conceptions (though they do). More fundamentally, global mobility gives rise to discrepant emplacements and conditions, whereby spaces and discourses sit uncannily in their surroundings, unreconciled or in negotiated tension with the sites and subjects with which they interact.

Discrepant emplacements: here we arrive at *heterotopia*, our central concept, the term around which this volume revolves. For what other name could grasp the clashing, incongruous spatiality of contemporary globalisation? No other "keyword" or conceptual rubric, we believe, simultaneously encompasses both globalisation's inherent difference and disjunctures, and its ceaseless production and puncturing of places. Hetero-topia: the word alone balances difference and spatiality in an incisive compound. In Michel Foucault's initial formulations, the term describes the spatial articulation of another discursive order. Heterotopias are discrepant segments of larger discursive totalities, whether those be a geographical or geopolitical space (such as, for example, a nation state) or the realm of language and textuality (such as a literary tradition). In relation to those larger bodies, heterotopias manifest their own distinct logics, moods, and norms. Heterotopian effects are also relational: they refract, disturb, but also accentuate aspects of the wider social or discursive totality. In their difference, heterotopias signal that another order, condition, or discourse can be realised, at least in part. At their most expansive, they are signs that a new social structure, a new way of life, is in the offing.

Heterotopia in the globalising present

Heterotopia and Globalisation in the Twenty-First Century revisits the theme of heterotopia in the globalising present. In so doing, we hope to bring a newly contemporary orientation, and a new empiricism, to critical writing on heterotopia. Much recent work on heterotopia has focused on the nineteenth and twentieth centuries, in which the onset of modernity brought on the emplacement, the concrete spatial instantiation, of new, humanly-confected regimes of rational efficiency, social order, and imperial authority. For example, Casarino (2002) discusses the nineteenth-century sea narrative as a "laboratory for the conceptualization of modernity" (9), with the ship as the intersection of modernity as crisis and heterotopias as "forms of representation that disturb and undermine representation" (15). Hetherington (1997, 17) explicitly makes a similar connection: "Modernity is defined by the spatial play between freedom and control, and this is found most clearly in spaces of alternate ordering, heterotopia." His examples include the Palais Royal in Paris in the 1780s, with its coffee houses as bourgeois spaces that played a role in the French Revolution and the emergence of modern society in France. This historical outlook remains prevalent to this day, as can be seen in Hancock, Faramelli, and White's volume *Spaces of Crisis and Critique: Heterotopias Beyond Foucault* (2018), which includes chapters that turn to Sade, Bataille, and Bachelard, again reproducing the retrospective orientation in getting to grips with heterotopia. The nexus of heterotopia and modernity has therefore been thoroughly addressed. Indeed, even the examples put forward by Foucault in his well-known typology of archetypical heterotopia – the ship, library, hospital, museum, bedroom, train, colony, the Oriental carpet, cemetery, boarding school, and military service – are, as he himself put it, "characteristic of western culture in the nineteenth century" (20).[1] The forms and functions assumed by heterotopia in the cultural present, however, remain unexplored.

In leaping free of the retrospective preoccupation with modernity that has so often characterised writing on heterotopia, this volume is resolutely oriented towards the twenty-first century. In place of Foucault's oft-repeated examples, the eleven contributions that make up the volume figure a new suite of heterotopian forms, sites, and relations. Twisting Foucault to our purposes, we might say these heterotopia are "characteristic of globalising cultures in the twenty-first century." Although each contributor attends to a concrete articulation of discourse, place, and practice, together the chapters chart a new round and patterning of heterotopia, which has largely displaced familiar examples. The chapters point towards a string of illustrative transitions: today, the colonial plantation has been transcended by agribusiness and soybean dictatorships, the honeymoon trip taken by train or liner by an Airbnb, booked and experienced online in advance. Fairgrounds, which formerly toured the edges of small towns, have been eclipsed by global countercultural festivals, the discrepant fragments of modernist literature by experimental digital films. The prison has morphed into immersive role-playing games set in former prisons, the natural sublime of the great national parks into the new ecological particularism of contemporary nature writing or the contested shorelines of climate and migration. These are, we recognise, just some of the heterotopias to have emerged or intensified in the twenty-first century. But they paint a picture. Heterotopias are changing in ways that speak to some of the most intractable problems and hopeful possibilities in the present. Consider how the boarding school, Foucault's example of a heterotopian rite of passage to maturity, has been overtaken by internment camps for child climate migrants. Or how the oceangoing ship – Foucault's "heterotopia *par excellence*" – might now be imagined instead as an interstellar vessel.

Still, our intention in this volume is not solely to update ready-to-hand examples of heterotopia, important though that is. Beyond this, we claim that engaging with new and emergent manifestations of heterotopia reveals some of the central dynamics of the contemporary world. This comes to fore if we recognise how the idea of heterotopia stands in tension, if not outright contradiction, with the other spatial rubric broached in this introduction: *globalisation*, perhaps the most widely (if often vacuously) used spatial term now in circulation. Whereas heterotopia essentially entails spatio-discursive difference, globalisation is widely taken as the moniker for spatio-discursive integration – that is, for incipient totality, "the widening, deepening and speeding up of worldwide interconnectedness" (Held et al. 1999, 2). Prevailing rhetorics of globalisation speak of interconnection and integration, even holism. Early acolytes of globalisation theory in the 1990s anticipated the arrival – indeed, the fledgling reality – of a levelled, borderless world, whether that be in the circulation of money through truly global, truly free markets, or in the decentred creativity and feedback loops of cyberutopia.[2] Before them, at the beginning of the twentieth century, socialist internationalists imagined a globe united in Communism, while European empires projected globalities of a darker kind.[3]

In the twenty-first century, for its part, globalisation would seem to present a tumultuous *un*differentiation of cultural spaces, in which formerly integral identities bleed into one another, diverse polities are commonly exposed to ecological risks, and sovereign territories fade amid shifting new configurations. In saying this, it is important not to exaggerate the particular novelty or significance of contemporary globalisation. Timothy Synder does well to caution us that historical cultures always tend to believe that their globalisation is special, that theirs is the first or only enduring globalisation to date.[4] The globalisation we have, we propose in contrast, should rather be seen as emerging from out of the remainders and ruins, fragments and failures of the many other globalisations that have gone before. The most pressing contemporary avatars of globalisation, we suggest, are *migration*, encompassing both labour mobility on a hitherto unknown scale and refugees fleeing draught, poverty, and warfare; *ecological transformations* brought on by carbon dioxide emissions, which, though all-too particular in their causes and consequences, form a global process; *digital technologies* that infiltrate and reconstitute social institutions, knowledge, and practices, such that human societies are newly calculable, manipulable, and dynamic; and a drawn out *hybridisation of the bordered state*, the integrity of which, though not simply annulled, is being increasingly compromised by ongoing neoliberal austerity across the developed world, the digital manipulation of national public spheres, and sub-state transgressions of sovereign territory.

Taken together, these globalising flows and planetary precarities could be thought of as flattening the spatio-discursive difference that defines heterotopia. Instantaneously spanning geographical distances, transgressing borders, and compromising the apparent homogeneity of once discrete cultural and territorial fields, the combined forces of globalisation seem to present a "runaway world," which, if hardly a stable integrated whole, is irreversibly interwoven and tied up with itself (Giddens 2003, title). Such conditions would seem to undermine the very possibility that difference might perdure in the contemporary world system – that any realm of discourse or cultural space might retain immunity and distinction amid global flow. Contemporary globalising processes, one might suppose, preclude heterotopia in advance.

But of course globalisation is far more complex than this. Heterotopia, we claim, can help us to see and make sense of that complexity. Indeed, the central argument put forward in this volume is that heterotopia constitutes a privileged vehicle for unearthing globalisation's dissonances and discrepancies. In their basic recognition of difference, heterotopias embody and accentuate the tensions, complexities, and dislocations with which globalising processes are riven. Dreams of global integration conjure away the heterotopian reality that what globalisation *actually* entails are confrontations and mismatches, displacements and partial inroads – what Arjun Appadurai has influentially described in terms of "disjuncture and difference" (1990, title). Indeed, some of the paradigmatic clashes in the long histories of globalisation – the Spanish Empire overtaking the Mexica world, enslaved West Africans displaced to sugar plantations in North America – involve chasms of cultural difference that even several hundred years of "shared history" have neither levelled nor reconciled. Nothing leads us to imagine that globalisation in the twenty-first century will proceed without such difference, whether of this abyssal sort or that of everyday disjunctures, such as that encountered in and around the gentrified coffee shop.

Still, heterotopian thought entails more than seeing heterogeneity where homogeneity ostensibly reigns. As well as being simply "different," heterotopias are also uncanny, ethically ambivalent emplacements that fail to straddle the gap between idealised visions of the good society and intractable worldly realities. So often, globalising discourses and movements are inspired by distinctively utopian visions of the world – indeed, the whole world – as it should be. Heterotopias are not opposed to such utopia; certainly they are not dystopia. Rather, they are complex, ethically dissonant spaces and orderings that are often produced in and through the realisation of utopian ideas. In them, visions of modernity or globalisation, though often utopian in conception, are diffused and refracted, complicated and compromised, as they are put into practice and run up against the world as it is. In the twenty-first century, for example, it is widely held that digital networks will unleash a dynamic and decentred new global creativity, while in the twentieth, it was thought that socialist internationalism would overcome artificial differences among peoples and polities. And yet the results, we know, are often online echo chambers of anger and anomie in the first instance, and impoverished regions and authoritarian zones of exception in the second. As worldly emplacements of ideal globalisations amid fraught realities, heterotopias bring the core contradictions and repressed antinomies of globalisations to the surface. They are twilight doubles of optimistic and unambiguous globalising rhetorics, ambiguous formations that perpetually eschew the clarity of the ideal and the simplicity of the absent.

In fact, the relations among heterotopia and globalisation run still deeper than this. Heterotopian thought may unearth the difference and ambivalence embedded in otherwise smooth imaginations and processes of globalisation. But this intervention is not launched from a site outside globalisation. Our argument is that heterotopian difference is immanent to globalisation itself, that heterotopias are a necessary and unavoidable aspect of globalisation. The point here, essentially, is that globalising processes concentrate difference and create disjunctures. Globalisation is often thought of as "bursting bubbles," a force that undermines once sealed and stable worlds, and thus annuls heterogeneity. We propose, in contrast, that globalisation can also be conceptualised as the gathering, imbrication, and setting-at-odds of once unconnected sites and cultures. As a result of this, what was once autonomous specificity re-emerges as heterotopian difference. Without a prior relationality, established by globalisation, there would be no heterotopia, only places and discourses, untrammelled by difference.

Hence, in preconditioning clashes, contacts, and contrasts among formerly distant discursive domains, globalisations constantly constitute heterotopia. It is impossible that this should be otherwise, for no globalisation has ever been so complete and one-sided as to erase totally the distinctiveness of one or all of the cultures and regions that it brings into contact. And conversely, for its part, globalisation is scarcely imaginable without heterotopia. Were there no distinct spaces and discourses for globalising processes to connect, conflate, or set in tension, then a smooth homogeneous globality would always already have been achieved and in place – no process, no "-isation," would be necessary, or indeed possible, under such circumstances: the global would be prior, unquestioned fact.

Difference persists and proliferates amid globalising flows. In this sense, globalisation *is* its heterotopias, *is* the production of discrepant worlds around the world. Globalisation and heterotopia, then, are two conjoined but fractions twins, neither of which can be wholly dissociated from the essential processes at work in the other. Now as before, these twins play out their tension, manifesting a rich variety of discursive spaces in the process. This antimony has taken on a distinctive character in the twenty-first century. During the long emergence of modernity, heterotopias largely arrived as frontier outposts at the vanguard of an incipient global economy (the factory, capitalist arcade, or oceangoing ship) or centrally ordered social world (the barracks, asylum, and prison). Globality and universality, in that period, were still in the offing, still to assert themselves fully against an established *ancien regime*. In the twenty-first century, the situation is different. Today we come in the wake of about fifty years of intense globalising projects and processes, and still more intense globalisation rhetoric. Under these circumstances, heterotopia are less the scouts arriving ahead of a globalisation to come, than the surfacing of repressed or overlooked difference, the working out of inner contradictions in globalisations that are already far advanced. Indeed, the recent emergence of diverse new heterotopias may be a signal that discourses of globalisation, pronounced with such confidence in the closing decades of the twentieth century, are now tired and threadbare, no longer able to sustain the optimistic pretence of integration or unity. Diverse separations and separatisms – Catalonia, British exit from the European Union, Crimea and eastern Ukraine, "America first" – interrupt the serenity, if not the sustainability, of institutions purposing cultural and economic confederation. Seemingly unifying processes – international climate protocols, cyberutopia, global "development" programmes – refract into the heterotopian complexities of unstable new microclimates, digital workhouses, manipulable echo chambers, and ever starker geographical divisions of labour and wealth.

Heterotopia: points of departure

In naming the complex spatio-discursive manifestations of globalisation's inconsistencies, wavering, and antinomies, heterotopia is a topic for our times. Still, showing how globalising phenomena including digital technologies, climate change, global political economy, and migration are giving rise to a heterotopian multiplicity of discursive spaces, entails revisiting, and to some extent reorienting, heterotopia studies. In raising the concept of heterotopia, critical writers often move through an identifiable series of introductory steps. Commonly, the first is to trace the conceptual origins to Foucault's writing, chiefly in his short text "Of Other Spaces" (1967)[5] and in the preface to *The Order of Things* (1966). Rather than recapitulate these texts, we refer to others who have done so excellently (e.g. Johnson 2006; many chapters in Dehaene and De Cauter 2008; Palladino and Miller 2015; Knight 2017). The second step is to point out that the concept remains underdeveloped in Foucault's writings. Reflecting critically on Foucault's "Of Other Spaces," it becomes clear that although his argument hinges on "some invisible but visibly operational difference" (Genocchio 1995, 38), a clear definition of spatial difference is sorely lacking. Another problem, as Saldanha (2008) elaborates in detail, lies in Foucault's outline of six "principles" of heterotopia with a clear list of examples. Together, these give the tempting appearance of a coherent package ready to be employed in other contexts. In this regard, Foucault's text bespeaks a structuralist concern with totality that does not sit well with the intended focus on otherness – and leads others to reproduce this "static, crypto-Hegelian dichotomisation of centre and margin" (Saldanha 2008, 2092). A third step is to note the broad variety of issues that have been analysed using the concept of heterotopia, which further extends Foucault's already wide-ranging list of examples. In recent years the concept has been employed to analyse hospitals (Street and Coleman 2012); nightlife and a subcultural music scene in Australia (Gallan 2015); graffiti in Greece (Zaimakis 2015); cruise ships (Rankin and Collins 2017); landscape architecture (Ebbensgaard 2017); early childhood policy initiatives in the UK (Barron and Taylor 2017); tea and British identity (Weygandt 2018), to name but a few.

Ironically for a concept targeting difference, these common steps amount to a remarkably uniform protocol. Although we would like to acknowledge these steps, in this volume we feel no need to reproduce them at length. We recognise the problems in Foucault's "Of Other Spaces" text, but our aim is not to resolve its oversights and contradictions. In fact, our first theoretical point of departure is that Foucault's work represents neither an arbiter nor a definitive authority on questions of spatial difference. Although Foucault's writings will continue to provide a shared point of reference for scholars laying out their take on heterotopia, if the literature on heterotopia shows anything, it is that the concept should be – and has been – developed beyond Foucault's writing. Hence, rather than retrospective clarification, our concern in this volume is to explore the possibilities offered by reconceptualising heterotopia in such a way that it can help us address the spatial, social, cultural, and political questions that a contemporary globalised world presents to us.

The second point of departure for this volume is that a heterotopia is not, ideologically speaking, unambiguously good or bad. Our approach is thereby conceptually in line with some of the literature on modernity referred to earlier. For example, Hetherington (1997, 13) locates heterotopia precisely in "the gap between freedom associated with the 'good place' and the invisible and all pervasive 'nowhere' and yet everywhere of social order." Likewise, for Casarino (2002, 17) the ship is the "heterotopia par excellence of Western civilization," which is also a representation of and tool for imperialism. This foregrounding of the complex political dynamics of heterotopian spaces departs from a portion of the literature on heterotopia that overemphasises Foucault's (2008, 17) suggestion of "counter-emplacements, a sort of effectively realised utopias in which the real emplacements, all the other real emplacements that can be found within culture, are simultaneously represented, contested and inverted." The opposition and contestation proposed here are all too easily reduced to a simple resistance to power, rendering heterotopia the space for (Marxist) critique. As Genocchio (1995, 39) compellingly argues, " 'Of Other Spaces' is invariably called up (within a simplistic 'for/against' model of conventional politics) to provide the basis for some 'alternative' strategy of spatial interpretation which might be applied to any 'real' place." While we certainly acknowledge the prominence of critique when taking on heterotopia, one only needs to invoke Foucault's own examples (2008, 21) of boarding schools, military service, and "the Puritan societies that the English had founded in America" to see the complex politics of difference, resistance, and oppression at work in heterotopian spaces. Or as Hetherington (1997, 24) puts it, "Spaces of resistance are also spaces of alternative modes of ordering; they have their own codes, rules and symbols and they generate their own relations of power."

Again, our opening example of a hipster coffee shop serves to underpin this point for everyday life in the globalised twenty-first century. One can easily critique such a space as the "landing gear" of global neoliberal gentrification, whose aesthetic appeal through difference works to displace those on the socio-economic margins. At the same time, it is also a space where a precaritised class of workers develop new forms of culture and representation, whose process of working in such a space is a departure from the strictures of cultural institutions, and whose turn to social media and other digital means engenders new social practices and relations. The concept of heterotopia entails that one should neither reductively paint our hipster coffee place as exclusively a spatial agent of neoliberal capitalism, nor merely celebrate it as a space of freedom for a MacBook-wielding creative class resisting the bureaucratic spatiality of the office building either. The discrepancy that lies at the core of heterotopian spaces extends to how the concept can be the political mobilised.

Our last conceptual point of departure is that while we did not wish to prescribe any specific understanding on heterotopia to our contributors, we do wish to emphasise two aspects: relationality and representationality. The fundamentally relational nature of heterotopia can be understood by Foucault's turn to a mirror in "Of Other Spaces":

> The mirror functions as a heterotopia in the respect that it renders this place that I occupy at the moment when I look at myself in the looking glass at once absolutely real, connected with all the space that surrounds it, and absolutely unreal, since, in order to be perceived, it has to pass through this virtual point, which is over there.
>
> (2008, 17)

The mirror shows two things simultaneously: an immediate reality and another order – unreal, virtual, utopian. A heterotopia projects a possibility, a glimpse, of another ideal order precisely on the basis of its embeddedness in the surrounding reality. Heterotopian spaces are therefore not static or encapsulated alternatives to a dominant spatiality, at one end of some oppositional logic. In pointing to the possibility of an emergent otherness, heterotopias incorporate precisely that which they are other to. Hence, as Foucault indicates, what matters is not so much that a space *is* a heterotopia but that it *functions* as one.[6]

The relationality of heterotopia is bound up with its representational nature. To understand this point, it is instructive to turn to Foucault's introduction of the term in the preface to *The Order of Things* (2002). Prompted by Borges' famous Chinese encyclopaedia, Foucault discusses heterotopia precisely in the context of the "non-place of language" (xviii) and argues that heterotopia "desiccate speech, stop words in their tracks, contest the very possibility of grammar at its source; they dissolve our myths and sterilize the lyricism of our sentences" (xix). Clearly, the concern here is initially linguistic, not spatial disruption. Yet through the figure of the table, with its double meaning as a material place for organising things and as a conceptual means of ordering ideas and data, Foucault (2002, xix) locates a point where "language has intersected space." The language, speech, grammar, and myth that Foucault speaks of in this passage should be understood as a concern with disruptions in discourse more broadly. The point of heterotopian discrepancy is a disruption in structural orders of thought, of representation, and of power. Putting together these elements from Foucault's two texts, then, heterotopia becomes a concept that foregrounds the entanglement of the spatial and the discursive.[7] For example, the holiday resort is a heterotopian disruption of spatial structures (demarcated site, usually abroad) but also a disruption of discursive orders (which regulate work, leisure, sexuality, etc.) replete with a host of representational practices (ranging from the pictures on TripAdvisor, sending postcards, to showing one's tanned skin once back home). The crux of heterotopia, therefore, is that it conceptually opens up the spatial as discursive as well as the discursive as spatial.

Chapter outline

Taking off from these shared points of departure, the eleven chapters that make up this volume explore heterotopian forms, spaces, and practices in the context of twenty-first century globalisation. Building upon work in anthropology, geography, cultural, media, and urban studies, the contributions update the established figures of heterotopia. In so doing, they show how globalisation has transformed heterotopian manifestations, in that digitisation, migration, and planetary environmental change are producing new discursive clashes at ever-greater scales, new technologies and spaces, and new modes of cultural practice. The carefully selected empirical case studies originate from different parts of the world involving a range of cultural processes and practices that reveal complexities in globalising processes unfolding today. Although not aiming at comprehensiveness, the new typology of contemporary heterotopias goes beyond Foucault's initial list of archetypical "other spaces." As such, it proposes a fresh account of eccentric, alternative visions of heterotopia adequate to the contemporary moment.

In the opening chapter Lieven De Cauter departs from the view on heterotopia as the space of play and ritual, a hieratic, sacred, liminal space of otherness, to reconsider this notion in light of the rediscovery of the commons. From this fresh vantage point, heterotopia can be redefined as the commoning of the (un)common, the sharing of the uncommon. De Cauter contends that heterotopias have a critical function in situations of crisis and are therefore crucial for the current age of disaster – the geological age of humans defined as the Anthropocene. As extraordinary space-times, heterotopias possess healing powers for communities in crisis brought about by wars, occupation, and other forces that threaten and destroy the social fabric. Such crises annihilate "normal" public space and render private space ridden with trauma, overcrowded and reallocated as a space for retreat and shelter. The refugee being a central figure at the end of the Anthropocene, sociocultural artistic practices and heterotopias of hospitality, De Cauter argues, have an important role to play in mitigating the devastating effects of geographical displacement and socio-political disenfranchisement. By evoking the work of Sandi Hilal and Alessandro Petti in Palestinian camps, as well as practices like Cinemaximiliaan in Brussels as his primal examples, he demonstrates how these new heterotopias become spaces of anthropological recovery from trauma, exile, and even the state of "illegality." Thus heterotopias have their pivotal function for the commons, allowing for such fundamental forms of community building, based on the sharing of the common uncommon. The practices of commoning find their deepest expression in heterotopian moments of hospitality.

In the next chapter, Cathy Elliott also uses Foucault's ideas about heterotopia in relation to the Antropocene, pondering the way contemporary literature deals with the theme of planetary climate change. Elliott turns to the genre of British New Nature Writing and presents a detailed reading of works by authors such as Helen MacDonald, Robert Macfarlane, and Richard Mabey. The New Nature Writing, she argues, stands in stark opposition to globally circulating dystopian narratives. With a focus on Foucault's distinction between the emptiness of the utopian "no-place" and the real, messy, interlinked concreteness of heterotopia, she shows how these popular books are deeply invested in narratives of global climate catastrophe. Their *temporalities* are markedly different from those privileged by dystopian (non-)fiction writing about environmental change. Elliott suggests that the new nature writing is, at its most interesting, radically at odds with linear climate change narratives that operate by imagining either a rush towards dystopia or fast progress towards utopian solutions. Both these temporalities operate without reference to *place*. In contrast, it is in the interruptive temporality of place that the new nature writing is most engaged and engaging. Viewed from this perspective, the New Nature Writing acquires a heterotopian quality in that it interrupts the urgency of a dystopian or utopian politics of global catastrophe.

Climate change is also central to Mary Gearey's contribution, which is devoted to the phenomenon of community activism. She sees this as a creative and collaborative response to the inequitable impacts of neoliberalism on humans and their local environments. Based on empirical work undertaken in three interconnected waterside villages along the River Adur valley in West Sussex, UK, the chapter argues that some of the most dynamic examples of heterotopia develop from an unexpected source – elderly retirees. Gearey describes diverse activist practices, such as forming and leading local flood action groups, which enable elder autonomy and reduce flood risk; fundraising and enacting riparian rewilding schemes to improve river environment biodiversity; challenging hegemonic local history narratives through polemic poetry and storytelling events; and campaigning to change land management processes to improve downstream soil and water quality. Defined by their life experiences and motivated by a desire to reassert agency by countering political disenfranchisement, the "heterotopian" elders deploy a variety of direct action techniques including entryism in local governance roles, social media campaigning, and public art performances. Defying simple categorisation and refusing to be rendered socially invisible by the politics of austerity, they help us to consider other possible sustainable futures.

In the next chapter Gladys Pierpauli and Mariano Turzi take us to a different geopolitical region, that of South America. Focusing on globalising processes in the field of agricultural production the authors contend that the current model of global agribusiness is reorganising national territorial boundaries, reasserting space and folding countries into a single regional structure: the "Soybean Republic." This structure, they argue, corresponds with Foucault's seminal interpretation of heterotopia as the spatial articulation of a discursive order. Understanding discourse as a contentious social narrative that is imposed by the powerful, the chapter unpicks the heterogeneous discursive construction of the Soybean Republic, which Pierpauli and Turzi accordingly recast as the Soybean Republics. Zooming in on the case of Paraguay, they show how the implementation of a global economic model effectively transforms the country's geographical and political space. They demonstrate how global agriculture has empowered transnational corporate actors and *brasiguayos* – Brazilian producers and their descendants who dominate soybean production and land ownership in Paraguay at the expense of economically dispossessed and socially disenfranchised peasants. Accordingly, within the dominant global discourses modern, hardworking, and integrated agricultural entrepreneurs are set in contrast with the ostensibly backward, lazy, and subsistence model of peasant family agriculture. The authors conclude that whereas liberal globalisation might be expected to homogenise agricultural space, global agribusiness has deepened heterotopian difference in the peripheral regions of the world. In the "agro-heterotopia" of the Soybean Republic, the integrating drive of global, liberal, and capitalist agriculture is simultaneously represented, contested, and inverted.

Like Pierpauli and Turzi, Adam Kildare Cottrel also sheds light on how space is affected by the forces of globalisation. Cottrel presents a detailed analysis of Pedro Costa's film *Ossos* (1997), in which the Portuguese director follows everyday life in Lisbon's Fontainhas district, a compact assemblage of improvised housing, home to migrant dwellers and immigrants of former Portuguese colonies. Through meticulously documenting the day-to-day struggles of Fontainhas' residents, Costa's film offers a glimpse into what heterotopia might mean in the context of globalisation by underscoring the disparity between local places and global logics of capitalist space. *Ossos* frames heterotopia with an eye toward what Foucault has called the "epoch of juxtaposition," where disparate groups of people are nestled close together but live lives increasingly defined by separation and isolation. Emphasising the simultaneity of the non-simultaneous, the film registers an irreconcilable world, where the Stone Age and the Cyber Era live side-by-side. It reveals how local places are vulnerable to globalising flows, which flatten spatial and temporal difference, eradicating heterotopic space and heterochronic time in the process. Cottrel's engagement with *Ossos* also allows him to explore aesthetic qualities of contemporary art cinema which has become increasingly invested in formal deceleration, defined by a serial use of long takes, stunted plot progression, and static cameras. The author argues that this aesthetic deceleration offers us an alternative temporal flow through which viewers can disengage from the accelerated experience of the globalising world. This alternative cinematic rhythm asks viewers to foster consideration for how abstract forces, such as speed, infiltrate local places, contouring organic notions of time and space for the appropriation and exploitation of capitalist expansion.

Henrietta Simson also finds value in contemporary art's ability to challenge globalised, neo-liberal paradigms. Building upon her artistic practice Simson establishes a definition of the heterotopia that contests entrenched notions of the landscape image and its imbrication with the perspective paradigm inherent to capitalist visual order. She argues that a radical reformulation of what landscape is or does is required for the Anthropocene era, a reformulation that opposes deeply engrained subject/object dichotomies and develops landscape from a background in front of which human narratives are enacted, to a co-subject, actively shaping how space and environment are constructed. The artworks Simson discusses take historical landscape forms and reconfigure them so that they are brought into the present as radical propositions for our changing visual and spatial regimes. The reformulation of landscapes of late medieval/early Renaissance painting as twenty-first-century artworks introduces difference to contemporary capitalist visuality. The landscape background as heterotopia becomes a discursive twenty-first-century object, around which questions pertaining to human subject/landscape object are called into question. Foucault's notion of the heterotopia – a place (topos) that can exist as a site of otherness (heteros) that connects and crosses over in a non-linear way through historical time – reconfigures the relationship between language and the world, enabling this reimagining, unfixing landscape from its mutuality with representation. The landscape heterotopia removes the human subject from its centre, facilitating a redistribution of meaning, suggesting possible configurations between landscape, image, and space.

In the next chapter it is littoral landscape that deserves Ursula Kluwick and Virginia Richter's particular attention. The authors suggest that in many ways, the beach is a heterotopia *par excellence*. It functions as a site of fantasy production, associated with playful practices and the limited temporal order of holidays. However, the beach is also interconnected with metropolitan spaces of capitalist production and hence not unaffected by global processes such as conflict over resources (oil, but also sand), the slow violence of global warming and pollution, and the unequal distribution of and access to wealth, security, and human rights. In the twenty-first century, the age of global tourism, the beach's contradictory status as an imagined refuge from the increasing uniformity, surveillance, and acceleration of everyday life, and as the overcrowded, faceless site experienced in package holidays all over the world, has further intensified. Even as some of the tourist beach's counter-spatial character is collapsing, however, other differences become visible. Littoral heterotopia segues into non-place, with tourists coming face to face with maintenance staff; illegal migrants selling fake watches and handbags; and, last but not least, refugees for whom beaches around the Mediterranean are interminable holding areas rather than places of safety. Kluwick and Richter draw on photography and contemporary fiction to explore the beach as a heterotopia of difference. Through an extensive analysis of Chris Cleave's novel *The Other Hand* they argue that in texts depicting the crossing of global trajectories at the beach, littoral space can segue from vacationscape into the state of exception.

Vacationscapes also appear in Elham Bahmanteymouri and Farzaneh Haghighi's chapter, this time in the form of Airbnb, specialised in offering tourism experiences and accommodation in private houses during holidays or other periods spent abroad. The authors argue that Airbnb can be conceived as a digital and above all ephemeral manifestation of heterotopia. Bahmanteymouri and Haghighi explore how Airbnb and its culture emerged in response to the financial crisis of 2008, making use of underutilised assets and contributing to the general entrepreneurialisation of society. In its disruptive aspect – e.g. with respect to local tax systems or the role of the local government in the rental housing market – Airbnb can be thought of as a heterotopia, lived and inhabited, though not as a space of resistance. Instead, with reference to Schumpeter's theory of creative destruction, the authors conceive of the sharing economy as a renewal of capitalism, in which Airbnb emerges as a space that challenges norms and relations, providing a glimpse of a metamorphosis of capitalism. However, viewed through a

Foucauldian-Lefebvrian lens, Airbnb should be thought of as a notably ephemeral heterotopia whose difference is flattened once norms are restored.

To continue the exploration of the twenty-first century's ephemeral spaces, in his chapter Peter Johnson explores how the pervasive accessibility of digital communication technologies has an impact on the idea of heterotopia. Johnson critically examines claims that some conceptions of the internet or certain online sites might be identified as heterotopia. Finding such claims problematic and suggestive, the author investigates ways in which heterotopia might be seen in the context of the interface between humans and technology. Taking up thoughts from Michel Serres, especially the philosopher's popular book *Thumbelina*, he argues for a repositioning of heterotopia in terms of an emerging digital environment. Finally, the chapter recalls Foucault's textual version of heterotopia as a means of breaking down customary alignments of thought which can now function as a stimulus for discovering creative ways to reimagine the global in our digital environment.

The possibilities of heterotopia for creativity and imagination also lie at the heart of Hanneke Stuit's contribution. Stuit invokes one of the most known Foucauldian examples of heterotopia, the prison, to uncover its new unexpected uses through the critical analysis of the real life game *Prison Escape*. Located in the disused panoptic penitentiary in Breda, the game provides a playground through which players can get close to incarceration without "actually" being locked up. The game combines a voyeuristic experience of the penitentiary as a heterotopia of deviation with a more "open-ended" form of heterotopia, which suggests a disruptive potential of the panoptic space inscribing itself both spatially and discursively on player bodies. On the one hand, the author argues, the scripts, narratives, and visuals used to set up the game's immersion tap into and confirm the heterotopia of deviation present in what seems to be a voyeuristic, globalised carceral imaginary. On the other hand, *Prison Escape*'s situated experience allows players to observe and explore their own embodied response to this imaginary through the effects of play. By analysing the role of the prison cell in the game the chapter focuses on the relations between the space of (role) play facilitated by the material impressions of the prison as playground, the globalised scripts projected on its interface by the individual player, and the panoptic power of the building itself viewed from within the cell. The tension between inscription as a universal concept of the body (heterotopia of deviation) and incorporation as individual experiences of embodiment (heterotopia as space of play) allows us to observe how *Prison Escape* feeds back critically into the globalised carceral imaginary.

The following chapter further develops the understanding of heterotopia as dramatic stages for the performance of contradictions. Informed by this idea, Graham St John's analysis of Burning Man – a city-scale participatory arts gathering in Nevada's Black Rock Desert – illustrates how paradox is the dynamic currency of this annual event. Based on qualitative research, the chapter adapts heterotopia with the goal of comprehending the hyper-liminal "other space" of Black Rock City, a frontier carnival with a threatened heteroclite ethos. St John traces the exchange between two archetypal figures – the *artist* and *tourist* – to interpolate the event's stature as a *participatory spectacle*. He directs our attention to the community's response to an undermining "culture of convenience" – a crisis exposing tensions between "participants" and "spectators." The author subsequently argues that the community has negotiated this crisis in art projects that dramatise paradox. Through the discussion of three project examples he demonstrates how this *redressive artopia* holds the potential to transform eventgoers and enable "Burner" identity by navigating boundaries separating the artist/tourist, producer/consumer, participant/spectator, self/other. This event-city case study evaluates heterotopology for the study of intentionally transformational events.

Lastly, in his afterword Kevin Hetherington takes off from the notion of an afterword as a paratext, a textual element that does and does not relate to the main body – in other words, a heterotopic text. He uses this textual space to situate Foucault's writing on heterotopia within his larger oeuvre, rooted in a second phase concerned with the relationships between the seen, the known, and their effects. Within Foucault's larger work, a connection can be drawn between the heterotopian ship and the ship of fools as discussed in *Madness and Civilisation*, as a real-and-imagined – i.e. heterotopic – site where discourses of the Renaissance are represented. Hetherington moves on to consider a contemporary version of a nexus of folly (of capitalism) and the future of civilisation (in light of climate change) for which the site is once again the entanglement between the sea-going ship (of climate activists) and a discursive space (of Twitter). The North Atlantic becomes a heterotopic space for different challenges to the crisis of globalisation, from opposing ends of the political spectrum. Though with a different outlook than Foucault, Hetherington concludes with the ship as the defining heterotopia for the twenty-first century in which glimpses of a possible future become clear – a future of large-scale migration and climate change.

Collectively, the contributions assembled here represent an initial attempt to move beyond the overly theoretical orientation of the literature on heterotopia. Instead of re-theorising heterotopia retrospectively, the authors engage concrete articulations of heterotopia in the cultural present. This constitutes a new approach, stressing how contemporary case studies help revisit, specify, and develop overarching theoretical accounts of heterotopia. The volume draws attention to how globalising processes are riven by heterotopian tension and complexities. By emphasising spatial and discursive difference, the concept of heterotopia complicates understandings of globalisation, which is often imagined as the homogenisation of cultures and spaces. Against simplistic visions that the world is becoming one, this book argues that digital technologies, climate change, migration, and other phenomena associated with globalisation are giving rise to a heterotopian multiplicity of discursive spaces, which overlap and clash with one another in contemporary culture.

Notes

1 Foucault's other examples belong to twentieth-century modernity (think of the motel and cinema) or to a variety of eras (these would be the mirror, honeymoon, festival, garden, carpet, fairgrounds, sauna, and brothel).

2 As examples of globalisation theory's epochal rhetorics and magical presumptions in its heyday, one might cite Frances Cairncross's *The Death of Distance* (2001), Thomas Friedman's *The World is Flat* (2005), Kenichi Ohmae's *The Borderless World* (1990); and – most famously – Francis Fukuyama's *The End of History and the Last Man* (1992). For a contemporaneous, clear-sighted critique of globalisation discourse, see Smith (1997, especially 10–14). This critique is glossed in Ferdinand, Villaescusa-Illán, and Peeren (2019, 21–22).

3 For a short genealogy of modern imaginations of globalisation and the global, see Ferdinand, Villaescusa-Illán, and Peeren (2019, 14–19). For a documentary that traces prevailing discourses of connectivity, see Curtis (2016).

4 True, perhaps Synder's rejoinder is more limited than this. He emphasises that in asserting the novelty of globalisation today, contemporary discourse overlooks how closely it resembles that which prevailed before the First World War. Once we recognise the "long failure of the first globalization" for what it was, however, it comes as little surprise that "our own globalization has contradictions, has opponents, that it generates – that it generates opposition, that it generates ideas of the far right, sometimes the far left, that are against it" (Quoted in Clendenin 2018, unpaginated). A longer historical view of the pile-up of past globalisations is intimated in Matthew A. Taylor's discussion of planetarity. "Our current world," he writes, "is the product of . . . failed global solutions" (2016, 116).

5 Several translations of "Des Espaces Autres" are current: the translation by Jay Miskowiec that appeared in *Diacritics* in 1986 (Foucault 1986) which is perhaps the most frequently cited translation; the translation by Robert Hurley that appeared in 1998 as "Different Spaces" (Foucault 1998); and the translation by Lieven de Cauter and Michiel Dehaene that appeared in their volume on heterotopia (Foucault 2008). All translations have their own advantages and problems. For this introduction we are using the 2008 translation, mainly because it explicitly frames the difficulties of Foucault's text and its translation. For the rest of this volume, however, we have not prescribed a particular translation to the contributors because we wish to avoid treating any translation as more "authoritative."

6 As Hetherington (1997, 8) puts it, "Heterotopia are places of Otherness, whose Otherness is established through a relationship of difference with other sites, such that their presence either provides an unsettling of spatial and social relations or an alternative representation of spatial and social relations."

7 See also Casarino (2002, 12) for whom heterotopias "come into being as the interference between representational and nonrepresentational practices" – a representational crisis through which he reads literary works for their dealing with modernity as crisis. In addition, see Cenzatti (2008), who also emphasises the representational nature of heterotopia, combining Foucault's concept with a Lefebvrian concern with spaces of representation.

Works cited

Appadurai, Arjun. 1990. "Disjuncture and Difference in the Global Cultural Economy." *Theory, Culture & Society* 7 (2/3): 295–310.

Barron, Ian, and Lisa Taylor. 2017. "Eating and Scraping Away at Practice with Two-year-olds." *Pedagogy, Culture & Society* 25 (4): 567–81.

Cairncross, Frances. 2001. *The Death of Distances: How the Communications Revolution is Changing Our Lives*. London: Texere Publishing.

Casarino, Cesare. 2002. *Modernity at Sea: Melville, Marx, Conrad in Crisis*. Minneapolis: University of Minnesota Press.

Cenzatti, Marco. 2008. "Heterotopias of Difference." In *Heterotopia and the City: Public Space in a Postcivil Society*, edited by Michiel Dehaene and Lieven De Cauter, 75–85. Abingdon: Routledge.

Clendenin, Dan. 2018. "Twenty Lessons on Tyranny – An Interview with Timothy Snyder." *Medium*, July 2. Accessed July 24, 2019. https://medium.com/@danclend/twenty-lessons-on-tyranny-an-interview-with-timothy-snyder-92f9b00be1b.

Curtis, Adam, dir. 2016. "Acid Flashback & A World Without Power." In *Hypernormalisation* (Television documentary). London: BBC.

Dehaene, Michiel, and Lieven De Cauter, eds. 2008. *Heterotopia and the City: Public Space in a Postcivil Society*. Abingdon: Routledge.

Ebbensgaard, Casper. 2017. " 'I Like the Sound of Falling Water, It's Calming': Engineering Sensory Experiences Through Landscape Architecture." *Cultural Geographies* 24 (3): 441–55.

Ferdinand, Simon, Irene Villaescusa-Illán, and Esther Peeren. 2019. "Introduction. Other Globes: Past and Peripheral Imaginations of Globalization." In *Other Globes: Past and Peripheral Imaginations of Globalization*, edited by Simon Ferdinand, Irene Villaescusa-Illán, and Esther Peeren. Palgrave studies in Globalization, Culture, and Society. Cham: Palgrave Macmillan.

Foucault, Michel. 2008 [1967]. "Of Other Spaces." Translated by Lieven De Cauter and Michiel Dehaene. In *Heterotopia and the City: Public Space in a Postcivil Society*, edited by Michiel Dehaene and Lieven De Cauter, 13–29. Abingdon: Routledge.

———. 2002 [1966]. *The Order of Things: An Archeology of the Human Sciences*. London: Routledge.

———. 1998 [1967]. "Different Spaces." Translated by Robert Hurley. In *Essential Works of Foucault 1954–1984, Volume II: Aesthetics, Method and Epistemology*, edited by James Faubion, 175–85. New York: The New Press.

———. 1986 [1967]. "Of Other Spaces," Translated by Jay Miskowiec. *Diacritics* 16 (1): 22–7.

Friedman, Thomas L. 2005. *The World Is Flat: A Brief History of the Twenty-First Century*. New York: Farrar, Straus and Giroux.

Fukuyama, Francis. 1992. *The End of History and the Last Man*. New York: Free Press.

Gallan, Ben. 2015. "Night Lives: Heterotopia, Youth Transitions and Cultural Infrastructure in the Urban Night." *Urban Studies* 52 (3): 555–70.

Genocchio, Benjamin. 1995. "Discourse, Discontinuity, Difference: The Question of 'Other' Spaces." In *Postmodern Cities and Spaces*, edited by Sophie Watson and Katherine Gibson, 35–46. Oxford: Blackwell.

Giddens, Anthony. 2003. *Runaway World: How Globalization Is Reshaping Our Lives*. New York: Routledge.

Hancock, David, Anthony Faramelli, and Robert White, eds. 2018. *Spaces of Crisis and Critique: Heterotopias Beyond Foucault*. London: Bloomsbury.

Held, David, Anthony McGrew, David Goldblatt, and Jonathan Perraton. 1999. *Global Transformations, Politics, Economics, and Culture*. Stanford: Stanford University Press.

Hetherington, Kevin. 1997. *Badlands of Modernity: Heterotopia and Social Ordering*. London: Routledge.

Johnson, Peter. 2006. "Unravelling Foucault's 'Different Spaces'." *History of the Human Sciences* 19 (4): 75–90.

Knight, Kelvin. 2017. "Placeless Places: Resolving the Paradox of Foucault's Heterotopia." *Textual Practice* 31 (1): 141–58.

Ohmae, Kenichi. 1990. *The Borderless World: Power and Strategy in the Interlinked Economy*. New York: Harper Business.

Ossos. 1997. DVD. Directed by Pedro Costa. New York, NY: The Criterion Collection.

Palladino, Mariangela, and John Miller. 2015. "Introduction." In *The Globalization of Space: Foucault and Heterotopia*, edited by Mariangela Palladino and John Miller, 1–12. Abingdon: Routledge.

Rankin, Jonathan, and Francis Collins. 2017. "Enclosing Difference and Disruption: Assemblage, Heterotopia and the Cruise Ship." *Social and Cultural Geography* 18 (2): 224–44.

Saldanha, Arun. 2008. "Heterotopia and Structuralism." *Environment and Planning A* 40 (9): 2080–96.

Smith, Paul. 1997. *Millennial Dreams: Contemporary Culture and Capital in the North*. London and New York: Verso.

Street, Alice, and Simon Coleman. 2012. "Introduction: Real and Imagined Spaces." *Space and Culture* 15 (1): 4–17.

Taylor, Matthew A. 2016. "At Land's End: Novel Spaces and the Limits of Planetarity." *Novel: A Forum on Fiction* 49 (1): 116.

Weygandt, Ariel. 2018. "The Cup of the Empire: Understanding British Identity Through Tea in Victorian Literature." In *Who Decides? Competing Narratives in Constructing Tastes, Consumption and Choice*, edited by Nina Namaste and Marta Nadales, 143–56. Leiden: Brill.

Zaimakis, Yiannis. 2015. "'Welcome to the Civilization of Fear': On Political Graffiti Heterotopias in Greece in Times of Crisis." *Visual Communication* 14 (4): 373–96.

2 Other spaces for the Anthropocene

Heterotopia as dis-closure of the (un)common

Lieven De Cauter

For Michiel,

in fond memory of one of the most joyful

intellectual collaborations of my life.

Introduction

The rediscovery of the commons (Ostrom 2015; Hardt and Negri 2009; Bollier and Helfrich 2011; Dardot and Laval 2014; De Angelis 2017 and many others) sheds a new light on the concept of heterotopia. Against the enclosures of the commons, primitive accumulation through appropriation and privatisation, heterotopias appear as what I would propose to call "dis-closures of the commons,"[1] or rather, as I will explain, dis-closures of the (un)common, the common uncommon. Dis-closure is both the necessary closure and protection on the one hand and opening up and sharing of the common on the other (De Cauter 2019). Disclosure is not only the sharing, the revelation of a secret, like in hacking, but can also mean closure for protection, like we do with monuments, to preserve them. Heterotopia, also according to Foucault's principles of heterotopology, exactly functions according to this dynamic of opening and closure. In all cultures the sharing of the uncommon happens in special, enclosed spaces, where otherness is given its place.

"Commoning," to use the crucial neologism that has become the core of the commons discourse to indicate the commons as practice, as a verb not a noun (Linebaugh 2008, 279), is nothing else than this practice of sharing which makes up the core of the commons: besides the community, the resources and the rules to govern that common, it is this praxis of sharing and self-organisation that defines the essence of any commons. The less organisational sharing of culture is also based on that sort of commoning, and exactly heterotopias have the function to share the uncommon, the strange, the weird, the forbidden, the other. Heterotopia can be redefined as the "commoning" of the uncommon.

If we just have a glance at the examples of Foucault (2008) we can intuitively grasp this redefinition: from nude beaches to saunas, which are radical revelations of nudity; to the temples where the sacred is kept, protected and revealed;

to the theatres and cinemas where the horrors of Oedipus and Medea and all their descendants are revealed and shared; not to forget to the cemeteries where the very common and totally uncommon reality of death is given a place – all are special places where this "otherness" is shared, "commoned." The first aim of this text is to make this redefinition of heterotopia from the vantage point of the idea of commons plausible.

The second aim concerns the urgent questions of our time: what can heterotopia mean in times of uncertainty, in "liquid modernity" (Bauman 2007) and more broadly in the Anthropocene, the geological era in which the human species became predominant, even at a geological level? Geologists are still debating when the beginning of the era should be situated (Demos 2017), but we might be living its end: the moment that humanity (or is it capitalism and should we speak of the "Capitalocene"? See Haraway 2016; Demos 2017) has become so predominant that the ecosystem is tipping into catastrophic mode and the system collapses in all sorts of unpredictable ways. Hence, the question of heterotopia today could be phrased as "other spaces for the Anthropocene." More precisely I will try to show how heterotopia can play a crucial role in the practices of commoning the uncommon in relation to one of the central figures of the Anthropocene: the refugee.

When we tried to clarify the concept of heterotopia for our book, Michiel Dehaene and I (2008) took the Foucault's text "Of Other Spaces" very seriously, almost word for word, and tried to really make a coherent concept out of it, in line with his six principles and in line with almost all of his examples. To have some firm ground to look back on heterotopia from the vantage point of the rediscovery of the commons and to be able to face the question of heterotopia in the Anthropocene, I start with a summary of our general theory of heterotopia (as we called it with some irony) as exposed in our book *Heterotopia and the City*, notably in our text "The Space of Play: Towards a General Theory of Heterotopia" (De Cauter and Dehaene 2008).

The third sphere: the general theory of heterotopia in a nutshell

Heterotopia literally means "other place" (*hetero-topos*). Foucault introduced the concept in a lecture for architects in 1967 but did nothing with it in his major works. The text did not appear until 1984, shortly before his death (Foucault 2008).[2] He himself regarded it clearly as a rough sketch. Foucault's examples give a good idea of the otherness of the places of alterity called heterotopia: the honeymoon, saunas, hammams, holiday villages, museums, theatres, cinemas, library, cemeteries, elderly homes, psychiatric institutions, gardens, parks, fairs, brothels – they are all heterotopias. What do they have in common? They are "different," "other." They interrupt the continuity of the everyday.

To outline these other places Foucault sketched six principles of what he called somewhat tongue in cheek a "heterotopology," a "science in the making." It is useful to briefly recall them here. The first principle is that all societies have heterotopias but that none are universal. So Foucault inscribes them as central in all societies. This is crucial because for a long time commentators had a tendency to situate heterotopia in the margins, on the threshold, even as invisible (see Hetherington 1997, 23–38; Dehaene and De Cauter 2008, 5–6). The second principle is that they can change function; for example, the cemetery was first the garden of the church where much of burial practice was collective, later shifting to the outskirts of the town and giving rise to the nineteenth-century pompous neoclassical cult of individual monumental graves. The third principle is that heterotopia can contain several spaces in one, the theatre scene being the key example here. It evokes the complexity of heterotopia in its mirroring and inverting relation to the ordinary world. The fourth principle states that heterotopias are linked to another time, that they are always also heterochronic. Here the extra-ordinary comes to the fore and points, as further discussed later, to rites of passage and liminal space. The fifth principle is closure, for all heterotopias have to be set apart as another space time. This principle is of course very relevant for my idea of dis-closure. Finally, all heterotopias have this mirroring/inverting function to the rest of society, theatre being a mirror of society and the city of the dead being our destiny even if opposed to the city of the living. It is clear that Foucault is outlining a sort of anthropology of extra-ordinary places. We should take this anthropological view of Foucault seriously. His heterotopology clearly points towards extra-ordinary places, set apart from the continuity of ordinary life. It is usually also a space that is not entirely public and not entirely private, but lies in between, mostly semi-public.

For our book Michiel Dehaene and I tried to make the concept more consistent, since it points to something very interesting but remained a little bit hazy. Because many of our concepts about space, economy and politics originate from antiquity and are derived from classical Greek, we thought it was imperative to return to the Greek *polis*. So, in our attempt to further define the contours of the concept of heterotopia, we started from Arendt's analysis in *The Human Condition* (1989) and along the way discovered a gem of a revelation in Aristotle. In addition to the sphere of the *oikos* (literally the household, both the private sphere and the private sector, the economy) and the sphere of the *agora* (literally the square, the public space, the space of politics), Hippodamus, an urbanist mentioned in the *Politika*, envisions a third space, which is neither economic nor political: the "hieratic," sacred or cultural space. In his utopia, the city (state) is divided into three domains: communal land (*koinèn*), private land (*idian*) and a part for the gods (*hieran*) (Aristotle 2005, 1267 b 30 – b 40). This tripartition can also be found in his city plan for the city of Milete: there are markets, agoras and "sanctuaries," sacred squares (which also contain hippodromes and theatres). We called this the discovery of the third space (De Cauter and Dehaene 2008, 90).

The holidays were called "hieromenia" (derived from *hieran*, holy). This time of the sacred is still clearly evident in the English word holiday: holy day. In other words, the third space of heterotopia corresponds to the sacred day, which interrupts and even suspends the everyday of economics and politics, very much in line with Foucault's principle of heterochronism. Heterotopia is therefore initially an event (in time) and only in the second instance a place, a building (a space). In other words, heterotopia is a time-space, a "sphere." We therefore called the collection of all heterotopias the third sphere (De Cauter and Dehaene 2008, 88–91), which consists of a cluster of heterotopias, other spaces, from the theatre, to the temple, the palaestra, the gymnasium, the stadium and the hippodrome or the bathhouse. The temple complex at Delphi could serve as a glorious visual example. Temporally we also see analogous clusters: during the *hieromenia* not only rituals took place and religious parades, but also games, in the form of competitions, such as horse competitions, athletics and music and theatre competitions.

Even in our time, where the distinction between the holy days and the ordinary days is becoming increasingly blurred, we still recognise the same contours. We still go to the temples of culture on Sundays, or during the weekend or on city trips, like we go to stadiums, concert halls, hammams, nudist beaches and cinemas – almost all of Foucault's examples.[3] The holidays are still the time for culture, sports, arts and religion. Heterotopia is the space of the cleansing, the feast, of the sacrifice and the ritual, of the gift, of the abundance and the wastefulness of the game, of representation, of the mimetic and the performative.

While we were working on our book *Heterotopia and the City*, at some point my eldest daughter burst in and told us about her Chrysostomos party. Chrysostomos is a high school ritual, a veritable rite of passage for high school students to mark the last 100 days of high school. In my daughter's school, the main event was a masked ball. To make her point, she said this phrase: "Whoever is not masked, cannot get in." This phrase was, for us, unforgettable because it marks in an utterly clear way the specific closure of heterotopia (the fourth principle of Foucault's heterotopology), and at the same time it contains the essence of the heterotopian in an almost graphic way. To be masked is to be other, masking is "othering." That is why children, who are by nature heterotopians, like it so much, because heterotopia is the realm of the mimetic and the performative. And because it is a party, it is of course also one of the highest practices of commoning: the feast, the festive. Almost all heterotopias are festive.

It is by giving the other, otherness, alterity, a place that heterotopia is socially (and therefore also politically) relevant. In short, religion, art, sports and games are all markedly social activities that we undertake outside the economic and the political, or phrased more cautiously: it is not the economic or political, but the anthropological (sacred and social, the common uncommon) that drives our activities in art, religion, sports and games. Work and labour are the characteristics of the private sphere (of the economy) and action is the nature of the public sphere (of politics), according to Arendt (1989). One could say that the activity of the third sphere, which Arendt overlooked somehow, is play or ritual behaviour. Huizinga's *Homo Ludens* (1952) was and remains a monumental contribution to this (re)discovery; like nobody before or after him, he laid bare the essentially playful dimension of human culture: an awful lot of human behaviour is game-like (and the game is a good metaphor to understand things human, think of Wittgenstein's language games). Heterotopia is the time-space of play in the broad sense: everything which is not work and not action. This is not only free time, but also ritual behaviour and study, as the term *skolè* indicates: school is essentially free time, time free of work and action. Likewise, artistic creation and performance take a central place in this ritual/playful realm of alterity. You see *plays* in the theatre and you *play* music. This heterotopian sphere, the secularised space of the "holy days," which we now call the culture, or the cultural sphere, is just as important as the economic or the political sphere but in a different way: as other space, as space of otherness.

In every society these spheres enter into a dialectic: heterotopia is a space of mediation between private and public, nature and culture, the realm of the living and the realm of the dead, between people and gods, between the rules of morality and the disturbances of desire. We took tragedy as an example. Women are invisible in the Greek *polis* in public and certainly in the political space of the agora, but on the scene of the theatre they can be represented as strong figures: Antigone, Medea, Elektra. In the third space a conflict can be brought up that cannot be shown on the public (political) space of *agora* or in the private space of the *oikos*. In Sophocles' *Antigone*, for example, the law of the *oikos*, "you will bury your dead," comes into conflict with the political regulation of Creon that her brother may not be buried because he has attacked Thebes and is therefore outlawed. In this tragedy, therefore, a conflict between private morality and public policy erupts. The conflict of Oedipus is also something that can be played out better on the stage than in the living room. But it does not always have to be tragic, it can also be comical, like in Aristophanes' *The Women's Parliament*. The main character Praxagora – literally she who acts (*praxis*) on the agora – violates the rules because she leaves the private sphere. She and her friends, dressed as men, assume control over the assembly in the early hours of the morning and immediately annul private property and declare free love. In Aristophanes' play, heterotopia becomes utopia and then degenerates into a comical dystopia, with people who do not pay their contribution, beautiful youngsters who have to make love with old ugly women, and more (De Cauter and Dehaene 2008, 93).

Heterotopia thus gives a place to that which has no place: from the taboos of nakedness, of Oedipus to the utopias of free love, from the numinous of the sacred and the mysteries of death to the gratuitous innocence of mere senseless play: sports. It is the "otherness," the extra-ordinary of heterotopia that represents and shows what else should be hidden. It is also the place of transition, of the body (sauna, brothel, motel) and eroticism. Foucault gives the example of the honeymoon; the honeymoon is the "elsewhere" of the defloration, the honeymoon is the "nowhere" (often a hotel) where *it* happens. This brings heterotopia close to the ritual of transition, as it was conceptualised by Arnold van Gennep (1981): all traditional societies have rituals of transition, from thresholds in space upon entering a house or a temple, to transitions in life (birth, adulthood, marriage, death). In his classic book *Les Rites de Passage* Van Gennep studied the unstable thresholds of these transitions which have to be stabilised by specific rituals. In the case of the honeymoon the transition from stable state 1 (virginity) to stable state two (being married) has to be performed through a ritual (the marriage), but there is an unstable transition space (the honeymoon) that situates the problematic transition outside the community, beyond the daily. "It" happens nowhere, as Foucault jokingly says. In 1967, the same year in which Foucault gave his heterotopia lecture, Victor Turner, directly inspired by Van Gennep, named this transition space *liminal* (from border, *limen*) if a real transition took place and *liminoid* if only an interruption of the daily takes place, for example a rock festival (see Faubion 2008 on this constellation between Foucault and Turner). Heterotopia is thus a threshold space or even a "borderline" space. Because it places the stable state of the everyday between brackets, it must also be demarcated and protected, by the playing surface, the boundaries, the seclusion, the limited access.

Therefore every heterotopia is a space set apart from the continuity of the normal by a system of thresholds and (dis)closures, to protect the space of ritual and play. During a game and a ritual the playing field or the temple is, in a sense, absolutely closed. Somebody entering a football field is in a sense just as much in transgression as a tourist during a holy mass: it interrupts the game and harms the ritual. *Temenos* means temple in Greek, but literally it simply means cut-out. Every temple (of culture) is a cut-out from the continuity of the everyday (De Cauter and Dehaene 2008, 95). It is this closure that is so important to set heterotopia apart from ordinary space/time, and this closure and opening determines, as we hope to show further down, the dis-closure of the (un)common in heterotopias, especially in times of crisis.

Our attempt to clarify and limit the concept of heterotopia had a clear program: against the experience economy, against the economisation of heterotopia – of museums, libraries, universities, art schools, theatres and other temples of culture which are all being reduced to enterprises – and against the political pressure to control them, these places of otherness in the *polis*, which are also places of deviation, dissent, eccentricity and excess, should be critically defended.

Heterotopian practices as the commoning of the (un)common

Against the background of our general theory of heterotopia, the rediscovery of the commons sheds a new light on heterotopia. What is shared in heterotopias basically is this special common: it is the commoning of the uncommon, the sharing of the meaningful and profound that we should but somehow cannot come to grips with. The common uncommon is that which is absolutely essential to the community and at the same time totally uncommon: nakedness, death and all other unspeakable things like the gods, the ultimate meaning of culture and life, etc. It is this function of heterotopia that I would like to call the dis-closure of the (un)common: like bare life, death, drama, the sublime, fear, the sacred. When this sounds farfetched and abstract Oedipus and Antigone make it concrete, but it could also be Van Gogh or Rothko. And almost all of Foucault's examples of heterotopias are the localisations of the unlocalisable, with nakedness and transience, love and death at their core. This dis-closure of the (un)common, the common uncommon is the function of ritual and play, of arts, sports and religion. But also the festive sharing of the group itself as the appearance of the common, communion and community as such. Of course the common(s) – of nature and culture – cannot be reduced to heterotopia, but I do believe that heterotopia is the place where the common appears as alterity, and alterity appears as common. Heterotopias are the places where this fascinating and threatening otherness – bare life and death and every other inexorable, inexplicable and/or important and unacceptable thing in between – is housed, tamed, adored, kept, defused. Both the cemetery and the theatre, the two only recurring examples in Foucault's text, are and remain excellent concrete examples of this fundamental anthropological operation.

The term of the "common uncommon," which Michiel Dehaene and I used intuitively and rather tongue in cheek, could be enriched by more recent uses. Blaser and de la Cadena (2017) argue that commons tend to overlook the conflict between the anthropocentric common good argued by governments to exploit resources and the commons as an environment including nonhumans, or even spirits, as might be the case in the approach of indigenous people to the same resource. For instance, the uncommon in approaching a river as a community living beside it, a source of hydraulic energy and a habitat of biodiversity, entails realising that different actors see a different commons (Blaser and de la Cadena 2017). They argue that we have to become aware of the "equivocations" (a term they borrow from Eduardo Viveiros de Castro) to fully understand the "uncommonalities" that are part and in fact constitutive of the commons. The authors also use the spelling "(un)common." The main thing we can learn from this unexpected and refreshing approach in anthropology is that equivocations and ambiguity are vital in all things cultural, most tangibly in art, which makes them both common and uncommon. For this reason art is both adored as the epitome of culture and despised as useless luxurious nonsense for the rich. That is the magic of art and its Achilles heel: it is just an ambiguous game with signs and images, which also explains why artists are descendants of shamans and medicine men. Some artists, like Dali, Warhol or Beuys, have taken up this role explicitly. As outsiders they deal with what really matters but they are also always charlatans and clowns showing a distorted mirror to society. Artists are inhabitants of heterotopia – like priests and theatre people, heterotopians par excellence – and because of that revered and despised.

Like the religious sacred, art is both attractive and intimidating. The sharing of art is similar to all other anthropological rituals surrounding the sacred, and it is barely a metaphor when we speak about "temples of culture." Even their often deplored "high thresholds" are not accidental. The ambiguous, open, playful and self-critical nature of modern and contemporary art has fuelled this distrust and has made the equivocations and the mismatches only worse. But all these ambiguities are essential: the fact that all "commoning of art" is always an unending process of interpretation points towards this cultural (un)common, similar to the iterations the "recursive ethnographer" does to deal with reciprocal translations of the uncommon inside the commons (Blaser and de la Cadena 2017, 190). Just like we have to decipher what we have in common when we see nature as a common cause from different perspectives and from different paradigms (scientific, activist, animistic), we have to share our impressions of films, theatre pieces and artworks to find out if we have actually seen the same thing. Heated exchanges are inevitable (like in sports, by the way). Interpretation is the life nerve of all things cultural, since language – a conventional system of ambiguous, "polyse-mic" signs – is the core of culture as human element.

Heterotopias are the laboratories, the breeding grounds of these ambiguities and equivocations, of this feast of mystery and meaning (ritual), of senselessness and gratuitousness of the rule itself (play). Therefore these places, all of them, since the dawn of civilisation to the last man, are crucial for anthropogenesis: it is in hetero-topias that we become human. It is the unending inculturations and acculturations into and out of the human element: culture; and the unending cleansing and medi-ating with our inhuman element: nature. In heterotopias culture and nature, spirit and body meet and mate. Even Aristotle thought high of dance in his *Poetica* and he ends his *Nicomachean Ethics* with musings on the softening effects of music.

Other support can be found in the work of Stavros Stavrides, who almost pains-takingly tries to solve the problem of closure in the commons. Indeed most com-mons in the sense of Ostrom are closed: you are a member of the community governing the irrigation system, the fishing ground or the alpine meadow or you are not; free-riding is fatal to those commons (Ostrom 2015). Stavrides (2016) tries to "expand the commons," by stressing "porosity" (a term he borrows from Walter Benjamin) of what he calls "shared heterotopias" (65–94):

> Heterotopias can be taken to concretise paradigmatic experiences of other-ness, defined by the porous and contested perimeter that separates normality from deviance. Because this perimeter is full of combining/separating thresh-olds, heterotopias are not simply places of the other, or the deviant as opposed to the normal, but places in which otherness proliferates, potentially spill-ing over into the neighbouring areas of "sameness." Heterotopias thus mark an osmosis between situated identities and experiences that can effectively destroy those strict taxonomies that ensure social reproduction. Through their osmotic boundaries, heterotopias diffuse a virus of change.
>
> (Stavrides 2016, 73)

A similar attempt to preserve a certain openness to the commons can be found in the book by De Angelis (2017). But at the same time a specific closure of het-erotopia protects the otherness and the other, and it is precisely this "apartness" that can make it a laboratory for change. This dialectic of osmotic porosity and protected apartness, we have called dis-closure. In contrast to the enclosures of the commons – the original appropriation by privatisations of the commons from the time of Morus to the present neoliberal wave of privatisations – we could call these practices of commoning the common uncommon: dis-closures of the (un)common.

Heterotopia in times of crisis: safe havens for the Anthropocene

Exactly because of such (dis)closures, I believe that in extreme situations hetero-topia can play a crucial role in a society and a community, to maintain a kind of gathering spaces and even to deal with trauma. It can help reshuffle the balance between the spheres, and can be an instance where a weak or non-existent public sphere needs to be strengthened or criticised. It can, in a sense, be a *stand-in* public sphere, without being immediately political, not exposed to the control of politics: apolitical or indirectly political maybe but far from party politics, policy and the police. As the public sphere came into existence in strange places like Palais Royal in Paris or the free mason lodges in England and elsewhere, as Hetherington (1997) brilliantly analysed, in a similar vein, the public sphere has been kept alive in all sorts of salons, bookshops, theatre and the like whenever freedom and democracy are under stress. It is in heterotopia, where artists and intellectuals flock together, that the critical spirit finds its origin and its hideout.

But here I want to focus on situations of distress, which characterises our current situation during this long ending of the Anthropocene, the moment that the predominance and footprint of the human race has become so massive that it is destroying the habitat of humans and non-humans alike, which Stengers and Latour have called "the intrusion of Gaia" (Stengers 2015; Latour 2015). Timothy Morton understands this as "dark ecology": *agrilogistics*, as he calls it, has brought upon us the split between nature and culture and this split has proved fatal, as extractivism and exploitation follow from this logic (Morton 2016). Migration is one of the many symptoms of this distress. In "liquid modernity" migrants are often considered as unwanted, surplus humanity (Bauman 2007).

Some years ago "illegal migrants," undocumented people or *sans papiers* occupied churches in Belgium to claim their rights. The Beguinage Church in Brussels, for instance, hosted several occupations between 1999 and 2014.[4] Because they were safe havens outside the private and public sphere, they had a different, ancient status of hospitality, of refuge. It is no accident that those who have no place to go, who are stripped of their rights (*bios*) and reduced to bare life (*zoē*), in the words of Agamben (1998), take to churches as protected ground, where the police cannot go, because in a sense the *temenos*, the "cut-out" of the holy place, lies outside the political sphere, the sphere of policing. It is equally no accident that some theatres and universities took over from the churches the hosting of protest occupations and hunger strikes by illegal migrants. They are not really private grounds, nor public in the sense of state-owned and controlled, and universities are age-old alien bodies ("corporations" in the old sense) in the city, with their own uniform, rituals, rules and freedoms; in that sense universities are heterotopias and should not forget their origins.[5]

This sort of refuge is the opposite of the camp as paradigm. The concentration camp has recently become a paradigmatic space of the extreme condition, of the localised state of emergency; one can refer to Agamben (1998) but also to Guantanamo. We cast heterotopia as the opposite of the camp, a safe haven against the state of emergency (De Cauter and Dehaene 2008, 97–98). It can oppose both hyper-economisation and hyper-politics of emergency, occupation, war, civil war. The camp is characterised by an implosion of the spheres: the absence of the public, reduction of the private (to mere life: *zoē*), the lack of place for the common. But also refugee camps struggle with weak differentiation of spheres: a lack of public spaces and cultural spaces.

What the refugee camp needs is maybe not so much "public space" along the lines of squares, open spaces and spatial representations of politics, but rather heterotopias. I give a few examples. The cultural centre Al Finieq in the Dheisheh refugee camp in the West Bank (Bethlehem) provides a colossal example of what heterotopia can mean for a community under stress: it is a bottom-up heterotopia. Built by the people from the camp, it supplies the people with a theatre, a library, a garden, a gym, after class study rooms, a concert and wedding hall, etc. It is a true civic centre of the camp and social gathering place, but it is also a reason for pride and collective self-respect. It is furthermore a laboratory for political education via culture and conviviality (Petti and Hilal 2018, 178, 275). Since it is situated outside official politics it can be political in an oblique, diagonal way: not the way of open confrontation in the political arena – which in the case of the occupied Palestinian territories is absent, in a sense – but in a contemplative, artistic, heterotopian way. As the Al Finieq centre proves, it can question "occupation" by the state of Israel, by short-circuiting or even transgressing it. Heterotopian practices are not activism, but playful transgressions, opening up possibilities. The practices of return which Sandi Hilal and Alessandro Petti have enacted in workshops and exhibitions are doing this. The exhibition, the museum space (Foucault's example of heterotopia as heterochronism) is a way to experiment with these returns.

Like the initiative to declare Dheisheh Camp UNESCO World Heritage is at the same time real and fictitious, the process itself is contrasted in exhibitions and in Petti and Hilal's book (2018) with the erased villages where the inhabitants of the camp who have dreamed of return for 70 years, come from (250–260 – the exhibition was on show in Van Abbe Museum in spring 2019). So besides stressing that this camp is a heritage of humanity, it also alters the fixed dream about "the right of return" of the inhabitants, symbolised by the keys of their houses they have kept for all those years, as if one day they would just go back and open the door. But the reality is that these villages were erased and often turned into natural reserves by the Israeli state, to prevent any possibility of return. The museum can host this double movement of heritage and return, which cannot be practiced, but it can be exhibited and debated, also with camp inhabitants and officials. In fact, as heterotopia is always several places in one (think of the theatre), it is ideal for practices of return to forbidden/inaccessible places.

Of course the "real" declaration of Dheisheh Camp as UNESCO World Heritage would go beyond these micropolitics, to become a hard political fact. Here heterotopia almost literally becomes the dis-closure of an (un)common: an inconvenient truth that neither the camp residents, the Israeli state nor the outsiders seem to be able to come to grips with, but that needs sharing all the same. Heterotopian places and practices contain and exercise scenarios for the "decolonisation of the mind," as was the initial battle cry, borrowed from Frantz Fanon, of the practice of Sandi and Alessandro (and Eyal Weizman) which was aptly called DAAR: "Decolonizing Architecture Art Residency" (2013; *Daar* also means hearth in Arabic and I will come back to this hospitality). Or at least: heterotopias supply a protected environment, a laboratory, a stand-in public sphere, in anticipation of a true political differentiation of the spheres. When the house is not a home and the agora is absent, or the school is lacking, heterotopia (the other space of culture and leisure) can be both a safe haven and a laboratory for hope.

A further example of heterotopia in times of distress would be "Campus in camps," another initiative by architects Alessandro Petti and Sandi Hilal, a university campus in this same Dheisheh camp. It is a heterotopian school for camp youngsters, opening up the world for them via discussion and art, and transforming the deadlock of frozen "permanent temporariness" of the Palestinian refugee camps.[6] That is in a sense what they try to do in all their work via practices of heterotopian conviviality (Petti and Hilal 2018). After their many projects in Palestinian refugee camps, they transposed "the living room" to Sweden: in Boden they organised a sort of bottom-up meeting place for refugees by refugees, based on the idea that hosting gives agency, reversing the role of guest, who should integrate, into host, who can add the culture where he is newcomer (Petti and Hilal 2018, 359–69). This transformation from guest (refugee) to host proves the healing power of practices of heterotopian commoning. The common meal is the utmost symbol and embodiment of the common and the utmost practice of commoning: hospitality is the sacred idea of sharing food with a stranger, to bring the uncommon (the stranger) into the common (the own community), by affirming the humanity and the need to eat and drink and share all we have.

This sort of micropolitics is an aim in itself, but can be paradigmatic for a larger approach of these problems. Indeed, just like a heterotopia can be the embryo of a public sphere, like Al Finieq in Dheisheh camp, or campus in camps, in a similar manner a heterotopia can be a stand-in for the home. I think initiatives like "the living room" are needed and practiced in many forms. I give the example of Cinemaximiliaan in Brussels, a cultural initiative working with refugees, often without papers. The asylum seekers or undocumented people involved speak about the "Cinemaximiliaan family," indicating that for them these cultural houses really function as temporary homes and as improvised kinship.

Cinemaximiliaan arose from screening films in the improvised refugee camp of the Maximilian Park at the Brussels North Station that caused much commotion in the summer of 2015, as a visible disgrace and symptom of the "refugee crisis" in Belgium. Besides blankets, tents and food, there was also a need for culture and conviviality. It was an excellent idea to show films because it brings a moment of relaxation, but at the same time it stimulates social life. The initiators, Gwendoline Lootens and Gawan Fagard, continued to show films after the park was evacuated, in asylum centres and later also at private houses, the so-called home screenings. In the meantime, the initiators moved to a large house that has become a real reception centre for "newcomers," not only a living community for exiles, but also a production house. Now these newcomers make films, which are tutored by big names like Bela Tàr and a bunch of local personalities from the film world (Hans Van Nuffel, Michael Roskam), in which they can often give shape to their traumatic stories. These films are already being screened at film festivals.[7] Through these productions they give these young people a voice, or even agency, empowerment and emancipation, chances of processing their traumas.

These young people without papers are also guided in pursuing education and are assisted with their files and papers. But above all they have a warm home and a place in Cinemaximiliaan to drop by and hang out. This is a laboratory for working together and living together across linguistic, cultural and religious boundaries (as I could see with my own eyes, as I was philosopher in residence there for a few months of personal exile). It is literally an exemplary, paradigmatic experiment, a heterotopian laboratory in globalisation, a breeding ground of a new global culture for the Anthropocene. Here the meaning of the formula of heterotopia as dis-closure of the uncommon, gets a new, more concrete ring: it is by opening up familiarity, convivial space and a network that the refugees, with their uncommon, alien, traumatic experience are brought into a new common and that the large community of white intellectuals, artists and cultural folk involved can in their turn get acquainted with a new cross-cultural richness. Disclosing a new common by overcoming the uncommon could be the magic formula of this kind of initiative.

These examples point to the power of cultural spaces, of heterotopias, to give a place to the common, without being private (economic) nor being public (state owned). Heterotopian politics is symbolic politics, diagonalising the private and the public. It reclaims the common (education, culture, memory) and gives also the private and the public a new place. Or it can be a space of preparing the ground for public space and a politics to come: a space of hope. Through cross-cultural hospitality and sharing these practices dis-close the uncommon of migration and displacement by turning it into a common, even the embodiment of commoning par excellence: the common meal as universal sign of the commons, from the Eucharist to the Chinese ideogram for common, which is exactly that: two separate hands eating from the same bowl (De Cauter 2013).

Other places for the Anthropocene (conclusion)

What is the relevance of heterotopia in the Anthropocene? All the examples I gave deserve more attention and explanation, but I have neither the space here nor is it my point: my intention was not to excavate cases, however paradigmatic. My point is exactly the plurality of them, for the reader could no doubt think of others. One could say that the book of Stavros Stravrides is one long attempt to "expand the commons" by studying "shared heterotopias" (Stavrides 2016), also in the face of the Greek crisis after 2008. Even if it expands the concept of heterotopia towards dissolution, the entire book tries to think this *dis-closure* of the urban commons.

All this proves the topicality of heterotopia for the Anthropocene. It is not difficult to see that heterotopias will be most important in the chaos that will happen in the unfolding of (this end of) the Anthropocene: during the long collapse, in the collision of climate change – and loss of biodiversity, the scarcity of water and other vital resources – the demographic explosion and migration, and the capitalist logic of growth and overshoot, there will be an immense need for these other places and places of otherness. We will need these heterotopias as laboratories to invent ways of dealing with the inevitable collapse of our system as the climate catastrophe is unfolding. In contrast to the enclosures all around us (the neoliberal capitalist logic, like the scandalous shock treatment of Greece) we should pay attention to the dis-closures of (un)common in and during the heterotopian practices of commoning. One could say that in the uncommon situations of the Anthropocene (with the refugee crisis as its almost allegorical symptom) heterotopias more than ever have the task to embody and make the common, often by simple gestures like common meals, hospitality and forming basic solidarity networks.

One could argue that heterotopia is too anthropological, too transhistorical, too archaic and fixed for this liquid modernity, for this (neo)cyberpunk age that will be the ending of the Anthropocene. But I on the contrary believe that places like the museum, the exhibition, the special school like Campus in Camps, cultural centres like Al Finieq, but also the special "houses" like "The living room" in Boden and Van Abbe museum, and Cinemaximiliaan and so many "other places" all around the world, will keep their role as *hetero-topias*: as protected safe havens where otherness and alterity are hosted and at home, where the uncommon is commoned, both in artistic practices, which always deal with the equivocations and ambiguities of the (un)common, and in basic practices of hospitality. As the places par excellence of sharing otherness heterotopias have an important, maybe irreplaceable role to play, particularly in the inhospitable landscape of the Anthropocene.

Notes

1 I give the example of our book *Heterotopia and the City* (Dehaene and De Cauter 2008). This book was made by myself and my friend and colleague Michiel Dehaene, collecting the texts of a colloquium on the theme. All authors were paid by their (public) institutions because participating in colloquia is what academics do, so they were more or less all paid by the taxpayer. We cleared the rights for the illustrations in the book and even designed the cover. So it was a free book so to speak. Then Routledge published it and sold it for €120 a copy. Of course, most of the copies were bought by university libraries, so this Routledge book was subsidised by the tax payer a second time, for most university libraries are publically funded. A good example of the so-called market logic in academia: enclosure by "primitive accumulation" or appropriation of public funds. Through this business plan Routledge made the book almost inaccessible for the common reader/buyer (and even for the editors – ten copies at half the price is still €600, quite an investment to hand out the book to friends and colleagues), until somebody just hacked it and made it available for free, so now anybody, including myself and my students, can use our book for free. (Just type "heterotopia and the city" in your search engine and it is all yours.) That for me is a fine and fitting example of hacking as dis-closure.

2 Most probably Foucault never corrected the transcript from tape. There is even a mistake in the text that survived into *Dits et Ecrits*: speaking in a sequence on the cemetery Foucault speaks about "le vent de la cité" – the wind of the city; in an attempt to translate this, we discovered it should of course be "le ventre de la cité" – the cemetery as "the belly of the city" (see Dehaene and De Cauter 2008, 19).

3 Except the prison of course, that was so to speak a Foucauldian lapse by the author who tried for once to stay clear of all the rest of his work. It has to be repeated that Foucault never called the panopticon, the hospital or the prison heterotopias in his classical works on these institutions.

4 See for example www.standaard.be/cnt/3r1pfvmc www.vrt.be/vrtnws/nl/2014/08/12/opnieuw_asielzoekersinbrusselsebegijnhofkerk-1-2058591/
5 In that tradition, academic freedom should supply intellectuals the freedom to speak up against the powers that be, without fear for repression. The "temples of culture" and the "temples of science" are supplying a free space that can and should assure the right to protest from sacred ground. This academic freedom is constantly under pressure.
6 www.campusincamps.ps/ (accessed February 2019)
7 www.cinemaximiliaan.org/ (accessed February 2019)

Works cited

Agamben, Giorgio. 2007. *Profanations*. Translated by Jeff Fort. New York: Zone Books.
———. 1998. *Homo Sacer: Sovereign Power and Bare Life*. Stanford, CA: Stanford University Press.
Arendt, Hannah. 1989. *The Human Condition*. Chicago, IL and London: The University of Chicago Press.
Aristotle. 2005. *Politics*. The Loeb Classical Library, Cambridge, MA, and London: Harvard University Press.
Bauman, Zygmunt. 2007. *Liquid Times: Living in an Age of Uncertainty*. Cambridge: Polity Press.
Blaser, Mario, and Marisol de la Cadena. 2017. "The Uncommons: An Introduction." *Anthropologica* 57: 185–93.
Bollier, David, and Silke Helfrich, eds. 2011. *The Wealth of the Commons*. Amherst, MA: Levellers Press.
Dardot, Pierre, and Christian Laval. 2014. *Commun, Essai sur la révolution au XXI siècle*. Paris: Editions La Découverte.
De Angelis, Massimo. 2017. *Omnia Sunt Communia: On the Commons and the Transformation to Postcapitalism*. London: Zed books.
De Cauter, Lieven. 2019. "Dis-closures of the Common(s)." In Lessons in Urgency, blog on *De Wereld Morgen*. https://www.dewereldmorgen.be/community/dis-closures-of-the-commons-proposal-for-a-new-concept/.
———. 2016. "Utopia Revisited." In *A Truly Golden Handbook, The Scholarly Quest for Utopia*, edited by Veerle Achten, Geert Bouckaert, and Erik Schokkaert, 534–45. Leuven: Leuven University Press.
———. 2013. "Common Places: Preliminary Notes on the (Spatial) Commons." In *On De Wereld Morgen*. Accessed June 2019. www.dewereldmorgen.be/community/common-places-preliminary-notes-on-the-spatial-commons/.
De Cauter, Lieven, and Michiel Dehaene. 2008. "The Space of Play: Towards a General Theory of Heterotopia." In *Heterotopia and the City: Public Space in a Postcivil Society*, edited by Michiel Dehaene and Lieven De Cauter, 87–102. Abingdon: Routledge.
Decolonizing Architecture Art Residency/ Petti, Alessandro, Sandy Hilal, and Eyal Weisman. 2013. *Architecture After Revolution*. Berlin: Sternberg Press.
Dehaene, Michiel, and Lieven De Cauter. 2008. "Introduction: Heterotopia in a Postcivil Society." In *Heterotopia and the City: Public Space in a Postcivil Society*, edited by Michiel Dehaene and Lieven De Cauter, 3–9. Abingdon: Routledge.
Demos, T. J. 2017. *Against the Anthropocene: Visual Culture and the Environment Today*. Berlin: Sternberg Press.
Faubion, James D. 2008. "Heterotopia: An Ecology." In *Heterotopia and the City: Public Space in a Postcivil Society*, edited by Michiel Dehaene and Lieven De Cauter, 31–40. Abingdon: Routledge.

Foucault, Michel. 2008 [1967]. "Of Other Spaces." In *Heterotopia and the City: Public Space in a Postcivil Society*, edited by Michiel Dehaene and Lieven De Cauter, 87–102. Abingdon: Routledge.

Haraway, Donna J. 2016. *Staying with the Trouble, Making Kin in the Chthulucene*. Durham and London: Duke University Press.

Hardt, Michael, and Antonio Negri. 2009. *Commonwealth*. Cambridge, MA and London: Harvard University Press.

Harvey, David. 2013. *Rebel Cities*. New York and London: Verso.

Hetherington, Kevin. 1997. *The Badlands of Modernity: Heterotopia and Social Ordering*. London: Routledge.

Huizinga, Johan. 1952 [1938]. *Homo Ludens, Proeve Eener Bepaling van Het Spel-element der Cultuur*. Haarlem: Tjeenk Willink en Zoon.

Latour, Bruno. 2015. *Face à Gaia, Huit Conférences sur le Nouveau Régime Climatique*. Paris: La Découverte.

Linebaugh, Peter. 2008. *The Magna Carta Manifesto: Liberties and Commons for All*. Berkeley, Los Angeles and London: University of California Press.

Morton, Timothy. 2016. *Dark Ecology: For a Logic of Future Coexistence*. New York: Columbia University Press.

Ostrom, Elinor. 2015 [1990]. *Governing the Commons*. Cambridge: Cambridge University Press.

Petti, Alessandro, and Sandy Hilal. 2018. *Permanent Temporariness*. Stockholm: Art and Theory Publishing.

Stavrides, Stavros. 2016. *Common Space: The City as Commons*. London: Zed Books.

Stengers, Isabelle. 2015. *In Catastrophic Times: Resisting the Coming Barbarism*. Translated by Andrew Goffey. London: Open Humanities Press.

Turner, Victor. 1982. *From Ritual to Theatre: The Human Seriousness of Play*. New York: PAJ Publications.

Van Gennep, Arnold. 1981 [1909]. *Les Rites de Passage*. Paris: Picard.

3 H is for heterotopia

Temporalities of the "British New Nature Writing"[1]

Cathy Elliott

> There is nothing before me now but wind and chalk and wheat.
>
> Nothing. The hawk rouses again and begins to preen her covert feathers. The running deer and the running hare. Legacies of trade and invasion, farming, hunting, settlement. Hares were introduced, it is thought, by the Romans. Fallow deer certainly were. Pheasants, too, brought in their burnished hordes from Asia Minor. The partridges possessing this ground were originally from France, and the ones I see here were hatched in game farm forced-air incubators. The squirrel on the sweet chestnut? North America. Rabbits? Medieval introductions. Felt, meat, fur, feather, from all corners. But possessing the ground, all the same.

The quotation is from Helen Macdonald's memoir of grief and falconry, *H is for Hawk* (2014, 263), which was a New York Times bestseller and won the Samuel Johnson award and Costa book of the year. This passage is emblematic of the recent rise of the "British new nature writing," typified by the work of Macdonald and such other bestselling authors as Richard Mabey and Robert Macfarlane. The genre has become something of a publishing sensation, particularly since *Granta*, "the magazine of new writing," published its 2008 edition, *The New Nature Writing* (Cowley 2008). Most bookshops now stock a shelf or even a full section of the British new nature writing; a literary prize, the *Wainwright Prize*, is entering its fifth year, whilst a new prize named after Nan Shepherd is being launched to reward talented new voices from under-represented groups writing in the genre; a well-attended three-day literary festival celebrating new nature writing was held in London in 2016; and new works are published regularly with no apparent diminution of interest.

This chapter discusses how we might analyse the "British new nature writing" through the lens of heterotopia. I start from the assumption that it is no coincidence that this flourishing of interest coincides with an increasingly intense circulation of dystopian narratives of ruined nature and planetary climate change. I argue that, at its best, this genre represents a heterotopian intervention in the politics of global climate disaster, problematising both utopian and dystopian narratives of the future that seemingly operate without reference to *place*. In contrast, the new nature writers are immersed in "small parcels of the world" that are also, in some sense, "the totality of the world" (Foucault 2008 [1967], 467).

I further argue that it is important to pay attention to the temporality of nature writers' engagement with place. As we observe in the earlier quotation, temporality is a constitutive element of "this ground," of concrete specific places, which always emerge from particular histories and the telling of them. Macdonald's description of the chalky ground on which she stands does not tell a standard linear history. Rather, we are invited to imagine a landscape in which the creatures (including humans) that inhabit it, as well as different times and their legacies, are piled up on top of one another in no particular order and following no progressive or regressive ordering principle. This disorderly, *ad hoc*, accidental temporality, I suggest, can be contrasted with both fearful dystopian narratives in which we are soon to face a speedy descent into fire and flood and utopian dreams of slick progress towards technological solutions on a planetary scale. Moreover, this interruption of standard dystopian and utopian stories about the future is political: engaging with and writing about nature in detailed ways resists the politics of complacency and apathy that too readily suffuse engagements with climate change.

To begin my discussion of the new nature writing as heterotopian, I will defend an approach to heterotopia that is analytical and political, rather than descriptive or taxonomic. My contention is that it is not particularly interesting to try to decide which places are more heterotopian than others. It is instead more useful to think about Foucault's writings on heterotopia as a set of tools that help us understand and analyse real places in ways that disrupt utopian reasoning. This is in keeping with the broader context of Foucault's work as an anti-utopian thinker. I then relate heterotopian analysis to ideas about time and temporality, focusing on Foucault's own notion of heterochronicity and contrasting this with the linear temporalities that animate utopian and dystopian thought. Next, I delineate the temporalities of what I call "dystopian (non-)fiction": the genre of popular literature that describes climate change in terms of an accelerating descent into apocalypse. I criticise this approach as ultimately depoliticising and compare it to the new nature writing's heterochronic temporalities. These works, I argue, embody a properly heterotopian analysis and make possible intimate *political* alliances between and within the human and non-human natural world.

Utopia, heterotopia, and heterochronicity

Much has been written about what Foucault meant by "heterotopia" and what sorts of spaces, relations, and orderings might count as heterotopian. However, I want to start from the first and fundamental distinction that he makes in the essay "Of Other Spaces": the distinction between utopia and heterotopia. It is important to place this distinction in the context of Foucault's other work. For Foucault, utopias are always "emplacements with no real place" (Foucault 2008, 394), recognisable by their own impossibility. As soon as they are realised in concrete form, they cease to be utopian in the sense of the "good place" and instead take on dystopian qualities, in the sense of utopia-gone-wrong. He is thus not only an anti-utopian thinker but also a remarkable analyst of the ways in which utopian thought produces certain consequences quite unintentionally. For example, and most famously, his work on the prison demonstrates the ways in which well-intentioned utopian thought, including fantasies of punishment without pain, delay, or uncertainty, and the sorts of spatial utopias that might eliminate crime such as the perfectly governed city or the panopticon, consistently fail in their stated aims but nevertheless instantiate modes of thought and practices of surveillance and discipline that perpetuate violence even as they hide it from sight (Foucault 1991, 11, 273, 198, 174, 205). This insight is not restricted to the prison but can also be seen for example in the "model town," which – when actually built – becomes a "whole series of disciplinary mechanisms in the working class estate" (Foucault 2003, 251). The continued success of the prison and the estate as ideas and aspirations – despite their failure to do the things they promise – are made possible through the "invincible utopias" whose designs and schemes continue to animate them (Foucault 2003, 271).

In "Of Other Spaces," Foucault goes on to describe heterotopia as "real places" and places that "really exist" (Foucault 2008, 404), signalling that this is "by way of contrast to utopias." It is therefore important to consider this proposition in light of his broader anti-utopian thought. If utopias are dangerous because they cannot exist and any attempt to bring them into being produces unintended and, often, malign consequences, heterotopias elude such dangers because they exist already. The utopian dream is to start again, as if from scratch, to knock down the slums and build a housing estate; to abolish existing systems of punishment and censure, and histories of motives and reasons, and produce the perfect new system for rehabilitation and new starts; or even to rewind and return to "bucolic [. . .] utopias" (Foucault 2003, 196) of the always imaginary past. This implies a new emplacement totally separable from what already exists, which must be razed to allow for the emplacement of the utopia. The heterotopia, meanwhile, is here already: we find it in a world already suffused with systems of power/knowledge, already messy, concrete, and fully entangled with other emplacements such that they cannot be separated. As such, the heterotopia always carries within it the possibility for resistance and reimagining, because there is no social totality, but rather myriad contending possibilities.

The literature on heterotopia is often preoccupied with making judgements about whether a particular place is, or is not, a heterotopia (for example, many of the chapters in Dehaene and De Cauter 2008; see also Johnson 2013 on the dizzying array of places that have been proposed as heterotopias). Intriguingly, Daan Wesselman (2013) departs from this method of categorisation in his discussion of the High Line park in Manhattan by comparing it to a short story "The Balloon" by Donald Barthelme, in which he discerns "a perfect literary instance of heterotopia" (25). He argues that although the eponymous balloon may have more "conceptual rigor" than the High Line, it is not useful to employ the concept of heterotopia to categorise or simplify real spaces such as the park. He suggests instead that we make use of heterotopia as an analytical tool (25). Wesselman wants to develop an understanding of the complexity of real spaces insofar as they depart from literary perfection and yet can still "engender otherness" as he puts it. The High Line, unlike the balloon, contains mundane elements of city life, notably an ice cream cart that punctures the otherworldliness of the place and reintroduces conformity to the values of consumerism into this space of "otherness." The High Line is therefore a concrete place that links together the rhythms of leisure and business, play and work, nature and culture, botanical and human, in ways that transform our understanding of the purpose and meaning of pathways through the city, but depart from the purity of the balloon.

This account is a helpful departure from arguments about whether a given space is heterotopian or not. It opens up the possibility that heterotopias are always multifaceted and, because they do not represent a social totality, may well uphold the status quo in certain ways, even as they challenge it in others. Where I depart from Wesselman, though, is in a methodological point that seems to me to have important consequences: his starting point is "to use the [High Line] park to critically reflect on Foucault's concept" (16). In order to do this, he elaborates the concept itself by using the balloon in the short story as a kind of pure exemplar. Because it is fictional, the balloon can really be nothing other than an alternative to the grid of the city streets. However, the reason that the balloon can attain the perfection or rigour that Wesselman attributes to it is precisely because, like Foucault's *utopias*, it does not exist. The High Line will of course fall short of the balloon's imaginative, but *imaginary*, perfection: that is part of the logic of a heterotopia.

Yet, by using the concept of heterotopia to foreground and celebrate the ways in which real spaces differ from utopias like the balloon, we could approach the High Line by taking seriously the idea that the ice cream cart is transformed by its emplacement in the other world of the High Line. The city streets below the High Line are frenetic and taken up with consumerism, and even here consumer purchases break up the impossible hope of an ideal otherness. However, whilst the vendor continues working in this other space, the consumer cannot: there is no immediate escape back into the normal routines of the city, and the ice cream must be consumed at a leisurely pace in the midst of the slower rhythms of the park.

Wesselman does well, then, to use heterotopia as a method of engaging critically with the spaces we already inhabit. However, by starting out with the *place* as a way of elaborating the *concept*, he perhaps downplays the ways in which the *concept* can be used as a way of understanding this, and any, real place. This, it seems to me, leads to the distracting use of fictional perfection as an exemplar. This approach also leads Wesselmaan simply to regret the ice cream cart rather than incorporating it into a fully heterotopian analysis of a complex space.

Peter Johnson suggests that attempts at categorisation, in which we try to work out which places are *really* heterotopian and, perhaps, the details in which they fall short, miss the playfulness of Foucault's original sets of examples and fail to take up the "scissors to cut" offered by heterotopian critique (2013, 793, 800). As should be clear from the preceding discussion, I follow Johnson in suggesting that it may be more productive to use the concept of heterotopia as an "initial conceptual method of analysis" (791). Heterotopia is a tool with which to understand and analyse the world as it is, rather than an ideal type *against which* we make value judgements about the value of a given place or space.

As utopias do not – and by definition cannot – exist, the whole of the messy concrete world we inhabit, in all its interconnectedness, can be understood heterotopically: it is simply a question of noticing and theorising in a heterotopian, rather than utopian, way. Kevin Hetherington (1997) has argued that heterotopias are not simply marginal places to be celebrated for their very marginality, for freedom and liberation from social ordering, because this ignores not only the ways in which margins are always a source of social ordering with their own sets of rules and codes, but also the ways in which forms of ordering at the putative centre are never uncontested or static. He suggests that heterotopia are rather the locus of alternative social ordering which can function as resistance to more hegemonic forms, but never operate in isolation from them. It follows from this that heterotopia offers modes of critique. I want to push this further by suggesting that heterotopia *as a concept* offers us possibilities precisely to critique utopian ordering and reasoning. Utopias, I want to show, are no-place and do not exist, whereas what *does* exist is *always* heterotopian, can never support utopian hubris, is never totally a good place, innocent of power. Heterotopia always carries with it traces of interconnectedness with all the messy, fragmented, and contradictory power relations that precede and surround it.

Johnson holds up Matthew Gandy's analysis of Abney Park cemetery as a particularly good example of heterotopian analysis in practice because of the ways it "challenges conventional readings of the natural world," particularly the binaries of nature/culture, urban/rural, and human/non-human (Johnson 2013; see Gandy 2012). Instead of recapitulating these binaries, Gandy focuses on the unlikely alliances that can form within and between them. Such heterotopian approaches are a useful way of understanding how rural and urban worlds intermesh and interact (Loughran 2016). As such, they challenge utopian ideas about temporality, rationality, and scientific progress inherited from Enlightenment and Romantic ways of thinking about nature and human beings.

Foucault's own description of heterotopia is insistent on the relationship between space and time, suggesting that heterotopias function fully when linked to "heterochronism": "a sort of absolute break with their traditional time" (Foucault 2008, 476). Heterotopias reconstitute the relationship between space and time and offer an alternative *temporal* as well as social ordering. Foucault describes the cemetery and the museum, where different times inhabit the same space and "time never ceases to pile up," or the festivity, in which time is fugitive, a time out of time. Attempts to use heterotopia as a critical mode of analysis have therefore been rightly insistent on the ways in which such places recode the relationship between space and time (Maier 2013). The High Line and Abney Park Cemetery are each heterotopian because the temporalities and rhythms of the natural and social world are imbricated within them. The cemetery stills time, as gardens do, as well as reminding us of human mortality and eternity (Gandy 2012, 7). Meanwhile, the High Line – as an old railway line and a place whose time has passed – is also separated from the city, such that visitors must by necessity take their time to saunter along (Wesselman 2013, 18).

Returning to the concretely anti-utopian uses of heterotopia, heterochronism can be understood as a method of critiquing utopian temporalities and their associated projects. All Foucault's examples disrupt the notions of time flowing forward that have animated the progressive temporalities of modernity and development that, in turn, have fuelled so many fantastical doomed projects and underpinned the mundane measurability of clock time that keeps factories and prisons ticking forward in their totalising regimes. They also cut against attempts to rewind to a past that seemed to exist before things went awry. However, despite these disruptions, the time of the heterotopia is also never truly ruptural in the way that animates slum clearances or model towns, because it infuses real places whose connections to real pasts and futures cannot be fully dislocated.

I am therefore taking heterotopia and heterochronism to be modes of anti-utopian analysis. This way of thinking engages with real places and times as a method of disrupting modes of social and temporal ordering that rely on the unreality of utopian thinking. In doing so, it offers a method of resisting and reimagining the consequences of utopianism. I have also already suggested the relevance of these ideas to encounters between the human and natural world, by showing how the binaries that animate conventional ways of thinking are placed in question by attending to the interconnectedness of both implied by heterotopian analysis.

How does this relate to the new nature writing? I will now suggest that this genre is heterotopian in its engagement with *dystopian* temporalities, offering us a method of critiquing them and a mode of resistance to the politics of resignation and technocracy that they instantiate.

Dystopias of climate catastrophe: the temporality of disaster

On the same shelves in bookshops as the new nature writing, we often find breathless dystopian (non-)fiction narratives about climate catastrophe. One particularly recent and popular example is *The Uninhabitable Earth* by David Wallace-Wells (2019), which has hit the bestseller lists in the UK and US and which I will use as an exemplary text here. Despite its high profile and claims to originality, it is neither saying anything particularly new nor doing so in a particularly novel way (beyond being readable and strikingly honest about the author's own foibles when it comes to recycling).[2] In fact, I suggest, the dystopian tendencies of this sort of writing penetrate a great deal of conventional thinking about climate.

Before I discuss and critique these accounts, it is worth stressing that I, of course, do not dispute the enormous problem that they diagnose. In common with both dystopian (non-)fiction and the new nature writers, I take on trust the overwhelming consensus of scientific opinion that human-made climate change constitutes a threat to human and non-human life and welfare beyond what we can perhaps imagine and that action to address it is required. It is not within the skillset of political scientists like me to question or confirm that scientific consensus. However, given that we accept the predictions about the likely consequences, it is our role to consider the politics of how such truth claims are made, what sort of action is likely to be politically possible, how alliances for such action might be forged, and how we might engage imaginatively with what might be required of us.

What I want to suggest here is that dystopian (non-)fiction may not help. Therefore, I now spell out three features of this type of writing that I find ultimately depoliticising. Although utopias and dystopias are not exactly the same, my discussion nevertheless owes a debt to Foucault's critiques of utopian thought, particularly his analyses of "utopias-gone-wrong," which is one definition of a dystopia.

First, like utopias, the dystopian worlds conjured by this genre of writing are unreal. When Wallace-Wells writes of a world in which it will be too hot to move around outside in the Tropics without dying and malarial mosquitos fly around Copenhagen, it is precisely the unreality of these predictions that lend them their drama (Wallace-Wells 2019, 26). Nevertheless, they are at a remove from our experience and seem scarcely believable. If we do believe them, as no doubt we must, it is on an intellectual level and evidently – as carbon does not cease to burn – not in our bodies which act in the world. Maybe we are too used to fictions on page and screen to register the visceral shock that such narratives seem to demand (Waldman 2018). Even when the events described are all too real, such as vivid descriptions of current climate events such as fire tsunamis, lethal mud slides, and deadly heat waves, there is something oddly unreal and displaced about the ways they are described. On just one page, chosen fairly haphazardly, we learn of the deaths in heat waves of a total of at least 58, 200 people in five countries, plus a series of extraordinary consequences including the circulation of dangerous conspiracy theories and the unrelenting consumption of oil to cool the lucky few who have air conditioning (Wallace-Wells 2019, 41). Differences between any of these places are smoothed over by the sense that they are all united by the wall of fire that is engulfing them. They are therefore maintained in their lack of specificity and concreteness as somehow unreal, even as the reality of these disasters is impressed upon us.

Second, the facts and figures, numbers, and locations of actual and impending catastrophes, are reeled off one after another so as to produce a sense of acceleration and headlong speed. Linear time is often criticised as a feature of such doomed utopian projects as colonialism, nationalism, progress, modernity, and development, with their emphasis on the steady forward movement of time as a force for improvement or expansion (Robinson 2017; Baraitser 2017; Closs Stephens 2013; Hom 2010; Hutchings 2015). However, the form of linear temporality that animates dystopian (non-)fiction is different in that it rushes headlong towards disaster. Wallace-Wells signals a change in "the way we imagine our own futures as we begin to perceive, all around us, an acceleration of history and the diminishing of possibility that acceleration brings" (2019, 35). There is something apocalyptic about these timescales, which are both quite short – the time we have to make changes is running out – and also unimaginably long – climate change is irreversible and its legacy will last for millennia. Moreover, linear time is sometimes characterised as being "empty" (Benjamin 1968; Anderson 1983; Bhabha 2012): I take this to mean at least partly that it is indifferent to *place*. Ideas about modernity and development imply a particular temporal-spatialisation in which the "West" advances further along in time, leaving its others behind in the past. However, the logic of this temporality also flattens out differences between places: everywhere will reach the same stage of modernity, progress, and development when time is fully played out. The temporality of dystopian (non-)fiction often follows a similar logic: although different from the always delayed and deferred temporality of development because it is played out at high speed, the oncoming fire and flood will mean catastrophe everywhere from the Arctic to the Tropics.

Third, as Methmann and Rothe have demonstrated in relation to similar sorts of apocalyptic narratives that circulate in intergovernmental climate change conferences, there is something deeply depoliticising about their logic (Methmann and Rothe 2013). For them, this is to do with the fatalism engendered in the face of an overwhelming, uniform, and speedily oncoming future that appears inevitable and casts nature itself as an invincible adversary of humankind. They suggest that the logic of this narrative is that rather than enabling exceptional measures, it instead produces a set of inadequate technocratic solutions around risk management and adaptation. We see similarly small-scale and depoliticised solutions being proposed in popular dystopian (non-)fiction, which frame appropriate responses to climate change in terms of switching off lights, flying less, or voting for "greener" candidates, as well as adaptation methods such as thinking twice before buying property in Florida (Romm 2016, xiii). As eco-critic Timothy Clark puts it, this is a peculiar derangement of scale: "[H]orrifying predictions of climate chaos [are followed by] injunctions not to overfill the kettle or leave appliances on standby" (2011, 136).

Individualised approaches like this fail to encompass both the deeply social and political nature of climate change, as well as the profound alterations in human life required to address it. They are also relentlessly anthropocentric. Human beings are at the centre of apocalyptic accounts, and the natural world is cast as an "Other" and even an enemy. It is humans alone who are understood to possess solutions, and those solutions that are proposed, such as geoengineering on a global scale, are utopian in their faith in technical scientific solutions to solve essentially political problems.

Even responses to the degradation of nature that seem to centre on the natural world – such as the rewilding enthusiastically advocated by George Monbiot (2013) – operate with a linear approach to time. In the case of rewilding, the idea is to "rewind" to a time before human interference in order to restore natural order. This approach not only misunderstands the dynamism of natural environments, in that it is not possible to preserve some putative pristine past environment (Thomas 2017). It also reinforces the notion that human interference is something somehow *not* natural, thus rearticulating the nature/human binary and echoing utopian fantasies of the bucolic past that contrast with the dystopian, thoroughly human, unnatural landscape of the present or future.

It is the imaginative intervention of the new nature writing that, I suggest, throws this kind of narrative into relief by demonstrating the possibilities held open by different temporalities and what Gandy refers to as "heterotopic alliances" (Gandy 2012).

Bringing nature back to life: the British new nature writing

Wallace-Wells writes provocatively,

> I may be in a minority when I say that the world could lose much of what we think of as "nature", as far as I cared, so long as we could go on living as we have in the world left behind. The problem is, we can't.

> (2019, 35)

This bald admission of anthropocentrism is entirely at odds with the anxious watching and occasional outpourings of grief that we observe in the new nature writing.

Richard Mabey, who was described by *The Times* as "Britain's greatest living nature writer" (Mabey 2008, 15), takes on anthropocentrism directly. He points out that the utilitarianism and self-interest of mainstream environmentalism is based on a "custodial relationship [. . .] intrinsically one of 'us' and 'them'" (Mabey 2008, 1645), which reproduces the very divisions between the human and natural world, and the relations of power, control, and extraction that drive environmental destruction in the first place. On his diagnosis, such approaches privilege "caretaking over caring" in ways that reposition the non-human world as the object of intervention, when it would be more logical to start from the behaviour of humans who are destroying it. He advocates, instead, an approach that takes the world to be already shared, in which humans and non-humans are always embedded in common processes, inextricably and unavoidably affecting and depending on one another.

This approach requires not only "diligent watchfulnesss" but also "cherishing rather than caretaking" (Mabey 2008, 1672). The natural world demands not our rational schemes but rather our passionate involvement in intimate alliances. When Mabey describes beech trees and bluebells as his "companion plants" (Mabey 2008, 2118), he is punningly drawing attention to the interconnectedness of human and non-human nature by drawing on the double meaning of horticultural "companions" that promote one another's biological flourishing, and a companion in the sense of fellowship. He here draws attention to ways in which the survival of one species is always premised on survival of an interconnected web. To repeat Foucault's phrase, "small parcels of the world" are "the totality of the world" (2008, 467) insofar as they emerge through processes documented by Mabey – weather systems, migration patterns, processes of hybridisation – that are both deeply localised and also entirely global. More than this, though, Mabey is signalling a relationship between the human and non-human that is at the scale of the "companionable" and intimate, a set of alliances that is perhaps more critical for human flourishing than for non-human nature. Companion plants are the ones without which one may well not survive.

Importantly, it is the temporality of this writing that constitutes the most striking break with dystopian (non-)fiction. In contrast with a unidirectional, linear, accelerating time, the quotation from Helen Macdonald's *H is for Hawk* at the start of this chapter recalls Foucault's description of the museum or cemetery: "[H]eterotopias in which time never ceases to pile up" (Foucault 2008, 476). Many different times are present (in both senses) in the same place. Foucault notes that heterotopias function only when "people find themselves in an absolute break with their traditional time" (Foucault 2008, 476). However, we need a heterotopian mode of analysis to grasp it where we see it. Soon after Macdonald describes the deer and their Roman introduction, she meets a couple she knows by sight and tells them what she has seen: "'Yes,' he says. 'Isn't it a relief that there're things still like that, a real bit of Old England still left, despite all these immigrants coming in?'" (Macdonald 2014, 264). The tired old teleologies of a nation's progress through "homogeneous empty time" re-assert themselves here, and Macdonald's narratorial and social unease bespeaks wariness of complicity. She shares in their love of the land but the politics of such love need to be disentangled from "chalk-mysticism [. . .] a kind of history that concerns itself with purity, a sense of deep time and blood-belonging" (Macdonald 2014, 260).

The text enacts this work of disentangling by counterposing the utopian history, with its fantasy temporality ("deep time") and its impossible ideas about purity with a detailed heterotopian account. A heterotopian analysis of the land requires serious involvement in place, including an understanding of the ways in which "all corners" are brought together through various temporary alliances between human and non-human nature and whose ongoing echoes outlive their immediate purpose. This pulsing temporality of the echo intersects with the accretive temporality of piling up to challenge the parochial, flat, uneventful time of local purity.

Even the narrative ordering of Macdonald's account is heterotopian. A utopian account might start from the problematic argument and soundly rebut it with information about the immigrant provenance of the deer, so that purity of the right answer can make its fullest impact. But the encounter comes after the land's complicated history has been unpicked, so that the remarks are already undercut before they are uttered. The encounter leaves the narrator sad and, then, fuming but for the reader there is a playful, comic undercurrent to the rambler's misfiring attempt to make common cause through recourse to national purity. The place itself pushes back against attempts to capture it in utopian terms.

The terrifying speed and acceleration of dystopian (non-)fiction is slowed down by these writers' patient involvement with the natural world and its rhythms. It is central to the enterprise of nature writing that authors stop to look closely, to name and describe the full range of life that they find, whether on Macdonald's chalk down, Mabey's woods, or the more urban landscapes of Alys Fowler's excursions into the Birmingham canal network (2017) or Rob Cowen's suburban edgelands (2016). It is not possible to sense a headlong rush towards disaster at the same time as taking a magnifying glass to hunt for slime mould on bark (Fowler 2017, 34) or patiently training a hawk. Concrete emplacements that are teeming with life return us to a sense of real places, connected to other real places, and militate against the flattening out of spatiality that puts the whole world on the same path to doom. Thus, engagement with climate change is detailed and specific, rooted in concern for specific species, ecosystems, and patterns, and fully engaged in the unendurable grief that their loss entails. Meanwhile, the distinctions between nature and culture or urban and rural that underpin narratives of industrialised progress polluting what had been a pristine, wild world are pulled apart by nature's never-ending refusal to recede. Old coal mines, viaducts, and canal bridges are colonised by the wild: "Nature always moves back in [. . . and . . .] the slimy, rubbish-strewn inner-city canals, with their tattered buddlejas and soot-leafed alders, can do the same" (Fowler 2017, 72). This is not a return to a pristine pre-human wild, but the practical adaptation of wild nature to a new set of circumstances that is *ad hoc*, provisional, and draws human and non-human nature into intimate alliance.

If heterochronicity inaugurates a break with traditional time, the temporalities of the "new nature writing" do not impose a new uniform temporality. Rather, various times co-exist, sometimes in tension with one another. Whilst the echoing time of animals introduced by humans who now seem entirely at home, intersects with a piling up of time, we also find moments of haunting, such as when Neil Ansell describes the annual return of bluebells in a landscape that no-one remembers ever being a woodland (Ansell 2012). These moments of return also constitute a pulsing and cyclical temporality: Tim Dee writes twelve chapters, each representing a month of the year, but episodes from many Januaries over the decades of his life appear in the relevant chapter (Dee 2010). This mode of writing makes sense when describing patterns of migration and interspersed sightings of birds, but it disrupts the familiar form of the ordinary linear autobiographical narrative. This multiplicity of temporalities, then, constitutes a heterochronicity that breaks with, disrupts, and unravels the conventional teleologies of progressive utopian *and* apocalyptic dystopian time.

One of Robert Macfarlane's most popular works is *The Lost Words*, co-authored with and illustrated by Jackie Morris (Macfarlane and Morris 2017). The book emerges from the observation he makes earlier in *Landmarks* (a Sunday Times Number One bestseller) (Macfarlane 2015) about the disappearance of words like *otter*, *newt*, *bluebell*, and *acorn* from the Oxford Children's Dictionary, to be replaced by, for example, *broadband*, *celebrity*, and *voicemail*. In response to this turn of events, Macfarlane and Morris produced a beautifully illustrated book of twenty of those "lost words," each accompanied by a "spell," or poem. The response was noteworthy: grassroots campaigners raised money for it to be placed in primary and special schools in all of Scotland and much of England and Wales; it is also used by charities working with groups such as dementia sufferers, domestic abuse survivors, refugees, and adult and child hospital patients; it has been used in school projects and co-opted by graffiti artists, adapted by dance, theatre, and music companies, and turned into songs by folk singers (Macfarlane 2019). If it is not a coincidence that the new nature writing emerges and becomes popular at precisely the moment of most acute ecological danger, then the chord struck by this book demonstrates some political and emotional responses to the losses meted out by climate change. The trauma involved in the loss of our relationships with the non-human natural world, including the loss of the everyday words we have to name it, is ultimately unendurable and alliances to combat it spring up as irrepressibly as weeds through concrete.

Ecocritic Richard Kerridge argues that there is always a tussle among three different strategies in writing about environmental change. Authors can engage in "pragmatic incremental working within mainstream culture, apocalyptic attempts to shock people into change, and formal experimentation that [. . .] attempts to change fundamental concepts" (2013, 354). New nature writers undoubtedly, at times, use all of these techniques, and are certainly not immune from drawing on their own sense of impending apocalypse from time to time. However, this taxonomy seems to miss the simple point that Wendell Berry, farmer and American nature writer, puts his finger on: if we are to take care of the natural world, we need to love it (Moyers 2013). Political theorist Jane Bennett puts it another way:

> The wager is that, to some small but irreducible extent, one must be enamored with existence and occasionally even enchanted in the face of it in order to be capable of donating some of one's scarce mortal resources to the service of others. [. . .] What's to love about an alienated existence on a dead planet?
> (Bennett 2001, 4)

Love requires a different temporality from the speed of the dystopia: taking care is also a question of taking time. For example, James Rebanks celebrates the sense of being "hefted," or belonging in some instinctive sense, to the Lake District farm that his family has farmed for centuries (2016, xi). He decries the modern norm that a good life always involves going somewhere different and engaging in endless travel (xiv). Instead, he writes in detail about the value of human life lived for many generations of the same family in one place, in deep, enduring, and long-lasting relationships with surrounding human and non-human nature. His account does not only disrupt the environmentally devastating narrative that a good life depends on cosmopolitan travel and an endless parade of novel experiences. It is also a subtle rebuke to the kinds of rewilders who would remove his sheep from the land, or leave them at the mercy of predators, in order to create forests and introduce wolves in an attempt to rewind to a putative pre-human past. Whilst Rebanks himself is committed to nature and wildlife, his account of the identities and commitments at stake in land management over centuries in the Lake District exposes the utopian fantasy of a pristine return to the wild as just that. The dislocation of farming communities from their connection with the land would be as violent as any of the utopian projects described by Foucault.

Conclusion: interrupting the Apocalypse

When I went to see *The Lost Words* transformed into folk songs, I realised that large numbers of people in the packed theatre were crying. The emotion crackling around the room during the extended standing ovation was overwhelming: both joyfully enchanted and heartbroken. As Foucault puts it, we need to pay attention to the concreteness of heterotopia because

> we do not live in a kind of void, in which we could place individuals and things. We do not live inside a void that could be coloured with diverse shades of light; we live inside a set of relations that delineate emplacements that cannot be equated or in any way superimposed.
>
> (Foucault 2008, 381–91)

The peril of writing about climate change as a speedy descent into dystopia is, I have argued in this chapter, that it draws on the politics of the void and therefore maintains a sense of unreality and disconnection. Taking Foucault seriously in considering the set of relations within which we live does not entail a withdrawal or denial of the problem, but rather the opposite: a serious engagement with how our intimate connections with certain small parcels of the world imply a broader non-anthropocentric entanglement. Reactions to *The Lost Words* and the contemporary flowering of the new nature writing signal that those intimate connections between the human and non-human world are already there, reproduced by heterotopian analyses like these accessible, loved, and popular books. By interrupting the dystopian narratives of climate change, the new nature writing interrupts the technocratic, apathetic, and helpless approach that dystopian narratives embody. I suggest that this genre of writing assists us in using heterotopia as a lens and allows us to interrupt narratives of destruction. I therefore want to conclude by proposing that the very use of heterotopia as a concept to understand the natural world helps us to develop strategies and approaches that work *with* our need and love for the non-human world. As such, heterotopia as a mode of understanding will enable us to build deep and intimate alliances that cannot be reconciled with linear temporalities, but instead link us in heterochronic times to the whole of the world.

Notes

1 I am very grateful to the editors of this volume for their encouragement and detailed, invaluable feedback. I also thank the participants at the "Time and Fear" workshop at the University of Sussex, June 7, 2017, for thoughts about and enthusiasm for a very early draft and Emily Robinson for her usual help, comments, ideas, and encouragement. Thank you to my student, Anna Ashford, for long conversations about heterotopias and utopias.
2 Wallace-Wells seems to have got quite lucky in book sales in comparison to other authors writing quite similar work, such as Joseph Romm (2016) or Mike Berners-Lee (2019). This is alongside a steady stream of increasingly dire warnings from journalists and activists which often take a similar tone. See, for just one example, George Monbiot (2016).

Works cited

Anderson, Benedict. 1983. *Imagined Communities*. London: Verso.
Ansell, Neil. 2012. *Deep Country*. London: Penguin.

Baraitser, Lisa. 2017. *Enduring Time*. London: Bloomsbury Academic.

Benjamin, Walter. 1968. "Theses on the Philosophy of History." In *Illuminations*, edited by Hannah Arendt, 253–64. New York: Schocken Books.

Bennett, Jane. 2001. *The Enchantment of Modern Life*. Princeton, NJ: Princeton University Press.

Berners-Lee, Mike. 2019. *There Is No Planet B: A Handbook for the Make or Break Years*. Cambridge: Cambridge University Press.

Bhabha, Homi K. 2012. *The Location of Culture*. London: Routledge.

Clark, Timothy. 2011. *The Cambridge Introduction to Literature and the Environment*. Cambridge: Cambridge University Press.

Closs Stephens, Angharad. 2013. *The Persistence of Nationalism: From Imagined Communities to Urban Encounters*. London: Routledge.

Cowen, Rob. 2016. *Common Ground*. London: Penguin.

Cowley, Jason, ed. 2008. *Granta 102: The New Nature Writing*. London: Granta Publications.

Dee, Tim. 2010. *The Running Sky*. London: Random House.

Dehaene, Michiel, and Lieven De Cauter, eds. 2008. *Heterotopia and the City*. Kindle ed. Abingdon: Routledge.

Foucault, Michel. 2008 [1967]. "Of Other Spaces." In *Heterotopia and the City*, Kindle ed., edited by Michiel Dehaene and Lieven De Cauter. Abingdon: Routledge.

———. 2003. *Society Must be Defended: Lectures at the Collège De France, 1975–76*. London: Allen Lane.

———. 1991. *Discipline and Punish: The Birth of the Prison*. Harmondsworth: Penguin.

Fowler, Alys. 2017. *Hidden Nature*. London: Hachette UK.

Gandy, Matthew. 2012. "Queer Ecology: Nature, Sexuality, and Heterotopic Alliances." *Environment and Planning D: Society and Space* 30 (4): 727–47.

Hetherington, Kevin. 1997. *The Badlands of Modernity: Heterotopia and Social Ordering*. London: Routledge.

Hom, Andrew R. 2010. "Hegemonic Metronome: The Ascendancy of Western Standard Time." *Review of International Studies* 36 (4): 1145–70.

Hutchings, Kimberly. 2015. *Time and World Politics*. Oxford: Oxford University Press.

Jackson, Patrick Thaddeus. 2010. *The Conduct of Inquiry in International Relations*. London: Routledge.

Johnson, Peter. 2013. "The Geographies of Heterotopia." *Geography Compass* 7 (11): 790–803.

Kerridge, R. 2013. "18 * Ecocriticism." *The Year's Work in Critical and Cultural Theory* 21 (1): 345–74.

Loughran, Kevin. 2016. "Imbricated Spaces." *Sociological Theory* 34 (4): 311–34.

Mabey, Richard. 2008. *Nature Cure*. Kindle ed. London: Random House.

Macdonald, Helen. 2014. *H Is for Hawk*. London: Vintage Books.

Macfarlane, Robert. 2019. "How the Lost Words Became Songs to Save the Countryside." *The Guardian,* January 26. Accessed May 20, 2019. www.theguardian.com/music/2019/jan/16/spell-songs-robert-macfarlane-the-lost-words-vanishing-nature-folk-musicians.

Macfarlane, Robert, and Jackie Morris. 2017. *The Lost Words*. London: Hamish Hamilton.

Maier, Harry O. 2013. "Soja's Thirdspace, Foucault's Heterotopia and De Certeau's Practice: Time-Space and Social Geography in Emergent Christianity." *Historical Social Research/Historische Sozialforschung* 38 (3): 76–92.

Methmann, Chris, and Delf Rothe. 2013. "Apocalypse Now! From Exceptional Rhetoric to Risk Management in Global Climate Politics." In *Interpretive Approaches to Global*

Climate Governance: (De)constructing the Greenhouse, edited by Chris Methmann, Delf Rothe, and Benjamin Stephan. London: Routledge.

Monbiot, George. 2016. "The 13 Crises Humanity Now Faces." *The Guardian*, November 25. Accessed 20 May 2019. www.theguardian.com/commentisfree/2016/nov/25/13-crises-we-face-trump-soil-loss-global-collapse.

Moyers, Bill. 2013. "Wendell Berry: Poet and Prophet." *HuffPost*, October 2. Accessed May 20, 2019. www.huffpost.com/entry/wendell-berry-poet-proph_b_4031836.

Rebanks, James. 2016. *The Shepherd's Life*. London: Penguin.

Robinson, Emily. 2017. *The Language of Progressive Politics in Modern Britain*. London: Routledge.

Romm, Joseph. 2016. *Climate Change: What Everyone Needs to Know*. Oxford: Oxford University Press.

Thomas, Chris D. 2017. *Inheritors of the Earth: How Nature Is Thriving in an Age of Extinction*. London: Penguin.

Waldman, Katy. 2018. "How Climate-Change Fiction, of 'Cli-Fi' Forces Us to Confront the Incipient Death of the Planet." *New Yorker*, November 9. Accessed May 20, 2019. www.newyorker.com/books/page-turner/how-climate-change-fiction-or-cli-fi-forces-us-to-confront-the-incipient-death-of-the-planet.

Wallace-Wells, David. 2019. *The Uninhabitable Earth: A Story of the Future*. London: Allen Lane.

Wesselman, Daan. 2013. "The High Line, 'the Balloon,' and Heterotopia." *Space and Culture* 16 (1): 16–27.

4 Disruptive elders

Enacting heterotopias of the riverbank

Mary Gearey

Introduction

Austerity politics within the UK has impacted significantly on the lives of its citizens over the past decade (O'Hara 2015; Collins 2016). Reduced public spending on health services, infrastructure maintenance, housing and social care has led to funding crises in these sectors. The impacts of these austerity measures have been particularly acute in rural communities (Milbourne 2016). Rural communities, compared with urban environments, have an older population cohort. Therefore austerity impacts disproportionately on rural elders (Milbourne 2015).

Alongside those who have lived in the countryside for much of their lives many rural elders are new to these spaces. Life stage migration out of urban areas is a recognised phenomenon; significant numbers of people in the UK move to rural spaces when their working lives are complete and they have officially retired. Many factors drive this geographic relocation. Stockdale and MacLeod (2013) note that many use relocation as the rationale to transition out of full-time work in a pre-retirement period, through setting up their own businesses which they continue on a fractional basis as they age, whereas Glasgow and Brown (2012) observe that certain rural spaces are seen as "good" spaces to age, either due to lively civic and social amenities or through the beauty of the local natural environment. Rural migration is often motivated by a desire to recast ones-self whilst in good physical health (Curry, Burholt, and Hennessy 2014, Glasgow and Brown 2012). Attached to this are hopes and expectations regarding connected ways of living, to other human and local environments (Gearey 2018). Munoz et al. (2014) argue that alongside a desire for a cleaner, healthier living environment with greater access to nature, there is an attendant desire to create new social networks and to establish novel or renewed post-work identities. Developing new hobbies, undertaking voluntary work and becoming involved in local civic activities are seen as entry routes for this. Often these active elder in-migrants provide rural communities with valuable, diverse skills sets and capacity building.

It can be argued that the "politics of austerity" (Featherstone et al. 2012) which stems directly from the 2008 global financial crisis, has impacted adversely both on extant rural elders and new incomers. Often this cohort remains pigeonholed as "vulnerable," with research focused on what has been denied to them in terms of support or access to services. Whilst this scholarship is urgently needed, what is denied to this study cohort is any sense of their own independence or agency. More research is needed to explore elder citizens' self-directed responses to austerity measures to understand how globalisation has impacted them in myriad ways. This chapter suggests that elders' experiences of environmental degradation and state withdrawal from public services have generated personal and political epiphanies which cannot be easily understood or defined within the existing vocabulary of citizen activism. Instead, this chapter suggests that these responses are idiomatic of a heterotopian impulse: a desire to enable other values, actions and behaviours to come to the fore, in opposition to mainstream, hegemonic, expectations. It is argued that these emergent gerontocratic heterotopian mindsets are a direct result of living through changes wrought by globalisation within their lifeworlds. These elder citizens disrupt the normative expectations of what people their age "do" through changing the physical and discursive spaces in which they live and commune. Here, the riverbank is the physical space which is the loci for these encounters, which confront stereotypes around ageing and project alternative ways of living and being together. Appreciating these elder "lifeworlds" (Edgar 2006), wherein political awareness has formed later in life (Guillemot and Price 2017; Walker 2010), enables a consideration of the importance of experience gained over a lifetime. Moreover, the crucial role of place connectivity in shaping identities comes to the fore (Tuan 1977; Proshansky 1978). Our immersion in landscape is central in shaping our sense of self in relation to the world (Schumacher and Gillingham 1979; Ingold 2011) though this relationship is influenced by underlying normative social values (Stokowski 2002). As a result, to understand relationships of place we must also understand the political, as well as physical, landscapes in which people are enmeshed.

This chapter addresses these issues by exploring how responses to austerity politics in rural UK communities have revealed a hitherto unexplored aspect, namely political activism by elders in direct response to a perceived withdrawal of the state from local governance. This activism is directed at environmental change, particularly with regard to water resources. The first aim of the chapter is to understand how gerontocratic performances and actions may shape a particular kind of community resilience that can be identified as heterotopian. The opening section of the chapter will outline a way of linking heterotopian activism with degrowth scholarship, and in particular empirical examples of nowtopianism (Carlsson 2008; Gearey and Ravenscroft 2019). These linkages draw our attention to the ways in which other forms of community building and political resistance occur in peripheral "different spaces" in the here and now.

The second, and related aim, is to explore how these various forms of gerontocratic networking enrich our understanding when considering presentations of heterotopian activism within the relevant literature. The chapter opens with a brief review on debates surrounding pluralities of meaning with regards to heterotopian performances. Considering the importance of temporality for these actions, both in response to the timelines of environmental change and to the nature of advancing age for these actors, the chapter suggests that contemporary anti-austerity "nowtopian" responses and actions which affirm activism to initiate change in the present can be conceived as a heterotopian practise, rather than a prefigurative politics which works towards utopian goals (Berneri 1950). Utilising data drawn from empirical fieldwork, focusing on community responses to water resources management issues within three UK waterside rural villages, the chapter uses the case study results to highlight forms of elder heterotopian practice, focused at the riverbank level, which refutes expectations around age and agency and disrupts dominant techniques of governance.

Utopias, heterotopias and nowtopias

Utopian thinking underpins all political economies (Mannheim 2013; Jameson 1979). This does not mean all political systems are utopian; rather the pragmatism of political compromise is sweetened with hopes of better future ways of living together. All societies are based upon systems of thought, whether democratic, centrally planned or totalitarian, which imagine a journey towards an ideal state, whatever form that might ultimately take (Arendt 1973). Valerie Brown (2015) notes a distinction between utopianism, and its normative unidirectionality and utopian thinking which embraces plural and heterogenous approaches to problem solving. This then enables an escape from the dualism of e/utopia as success and dystopia as failure, to enable pathways which accept a movement towards collective attempts to overcome shortfalls and failures. Brown acknowledges Isiah Berlin's focus on utopian thinking as a creative, collaborative process, describing it as "hopeful and imaginative" (Brown 2015, 213).

Heterotopian analysis goes some way towards this. As explored throughout this book heterotopian scholarship takes its antecedence from Foucault's 1986 translated work "Of Other Spaces." Foucault's essay was exploratory, with most of the resulting scholarship coming from critiques and recombinations of elements of this work (Soja 1995; Lefevbre 1991). Heterotopian thinking asks us to consider space as an artefact of the society that produces it, such that a physical space can have multiple uses and signify different things to those in that space at any given moment. "Deviant" spaces are heterogeneous and abundant. Foucault used the rest home, the brothel and the playground (Johnson 2006) as examples of spaces of alternative thinking, practice and imagination to enact other possible selves. We can extend this to include the domestic, quasi-civic and public spaces where people talk, think, play and imagine together. Agency becomes diffused; heterotopianism does not just reside within the self-aware deviant, but also within those whose hopes, dreams, actions and ways of being run counter to that of the dominant hegemonic paradigm. There are slippages, and shape shifting as actions and responses can be interpreted and understood in multiple ways over different time cycles.

This is particularly pertinent as we focus on the importance of changing lives in rural communities within this chapter. Kevin Hetherington's 2002 work *The Badlands of Modernity* forwarded the idea that alongside the dominant social ordering of space in nascent modern communities were other civic or quasi-civic spaces within which counter-hegemonic ideas, actions and performances were enacted. The factory floor, the literary salon, the court-house and the laundry yard are all othering spaces in which (at least in the global North) modernity forms, and which in the late modern age would include our digital selves and communities. This calls our attention to the importance of marginal spaces, and alternative impulses or responses, in responding to, or disrupting, normative social and cultural practices in mainstream society. Drawing on Foucault, Peter Johnson argues that "whereas utopias are unreal, heterotopias are 'actually localizable'" (2013, 791) in ways that can fragment the linear trajectories of "development" in order to focus on interstitial moments and performative actions which reclaim power and agency in the "spaces in-between" (Foucault 1986). Johnson goes further to examine how the agile and disturbing potential of heterotopian sites challenges the dominant reading of them, unveiling a liberational potential. He contends: "Heterotopias are defined as sites which are embedded in aspects and stages of our lives and which somehow mirror and at the same time distort, unsettle or invert other spaces" (Johnson 2013, 791).

It is this empowering element of the heterotopian mindset – the urge to imagine and to make possible other futures – that is so crucial to nowtopian scholarship. Nowtopia is a novel term used to capture the different practices, actions and communities that are unified through a desire to demonstrate that capitalism is not the only form of collective being-in-the-world. In some respects it is a form of pre-figurative politics in which the politics that is being imagined is kindness. It is an affirmatory rejection of capitalism and of its globalised tentacles. Nowtopianism as an avant garde innovation in living together accords with degrowth theory; prosperity for all decoupled from continual economic development, consuming less, treating ourselves and other beings with dignity (D'Alisa et al. 2014). Working in tandem with critiques and analyses which seek to make explicit why capitalist and neoliberal strategies of accumulation are ultimately negative for human and other planetary lifeforms, it is not enough to uncover what's failing. There also have to be suggestions, or glimpses, of other ways of being. Nowtopianism in many respects aligns with this heterotopian purview to validate other ways of using and making space. Although a recognised component of the degrowth vocabulary (see D'Alisa et al. 2014), nowtopia as a term is nascent in its use and is attributed to the authors Chris Carlsson and Francesca Manning (2010). Carlsson and Manning define nowtopian thinking as "a term that attempts to describe the myriad efforts to reclaim and reinvent the meaning of human labour against the logic of capital" (2010, 925). From this perspective human work is presented as worthwhile, fulfilling and necessary. Nowtopianism doesn't offer a manifesto; nor is it a social movement. Instead its focus is to recognise the small scale, self-directed activities which seek to harness the value that is involved with labour but which itself is intentionally disconnected primarily from monetary value or ends. It is labour which supports community building, environmental care, self-esteem, creativity and social connectedness and, amongst innumerable other benefits, is working towards the nowtopian vision to unshackle labour from capitalist relationships and provide a common ground for human expression and interaction.

For Carlsson, those who follow this approach might be described as nowtopian entrepreneurs and adventurers who display the characteristics of "tinkerers, inventors and improvisational spirits" (2008 186). In the nowtopian world there are, thus, no hierarchies of the type so necessary to capitalist accumulation, and no subjugation of certain forms or ranks of labour in favour of others. As we consider what this might mean for a twenty-first-century reflection on heterotopian practices we can argue that these spaces of difference now are relocated in disparate physical spaces, i.e. not just within the city but also in rural spaces. They are also digitised, informatic spaces used by online communities with free and open access to these resources.

It is this rejection of hierarchies, this "commoning," which is explicitly unique to the nowtopian discourse. There is no idealised model for human social organisation; so in many ways its natural bedfellow is a Marxist appraisal of alternative models for collective human fulfilment. Put crudely, within Marx's work human nature is a *tabula rasa*, moulded by the historical-economic conditions of any given society. Marx speculated that the disenfranchisement of humans from their potential creative selves is driven by waged labour relations in our current Capitalocene (Moore 2017). For Marx, humans are imaginative, social creatures who create and live together with others. Production, creation and shared endeavours are elements of what it means to be a human; this worldview is shared by nowtopian practitioners. Labour is not the issue; waged labour within a capitalist paradigm is the issue. Unlike intentional utopias, nowtopia writing recognises the possibility, even the desire, for self-determined, unstructured and unplanned synergies of human activity which reclaim social relationships outside of monetised environments. In other words, nowtopias are all around us – they inhabit a new realm of heterotopian space – although we need new tools and vocabulary to identify them.

Yet, there is a tendency within the literature to identify the nowtopian drive or "impulse" as that of a working class struggle to reclaim agency against the strictures of capitalism. Thus, Carlsson and Manning contend that "Nowtopians are a part of the working class with a specific experience of capital, whose struggle, if cognizant of its resistance to capitalism, can feasibly link with other struggles over a common enemy" (2010, 930). This creates a boundary regarding who can be a nowtopian – surely the opposite of the open and inclusive approach that defines this worldview. There are then, presumptions which underpin nowtopian framings – that nowtopianism is a grassroots initiative of the working class, deliberately pitted against capitalist labour relations. Yet, it can be said that this simply replicates the binary producer-owner categorisation of capitalism, while simultaneously denying agency to those whose work practices define them as other than working class. Instead, this chapter argues that nowtopian individuals can remake themselves as part of a new other-working classlessness – not by othering themselves in some form of denial, but by creating new practices undertaken in new ways. This enriches our understanding of what contemporary modes of the heterotopian imaginary may look like. The agency is located in the practice of world-building, of demonstrating that other ways of being and thriving together are possible. As a result, it becomes possible to conceive of a more integrated and inclusive perspective of who and what nowtopianism can be, and how using a heterotopian critique of people, action and space can support this.

D'Alisa et al. have described the degrowth rationale when implemented as the "reproductive economy of care" (2014, 32). This gets to the heart of nowtopianism: the creating and enabling of forms of living, working and producing together which sits outside of capital exchange and instead generates new commons, new forms of relationality. Instead of the focus of much current degrowth scholarship on forms of nowtopian community building that are deliberately underpinned by an overtly political rationale, the chapter proposes that such intentionality is not inherent to all examples of nowtopias. Heterotopic practices or "episodes" can necessarily take multiple forms. These can be co-ordinated through social media affiliations, online group forums and group messaging on mobile phone apps, or be the product of unplanned, uncoordinated meetings and information sharing – at community events, festivals, school gates, garden centres – anywhere. These social movement connectivities are both highly visible through public interactions and underground through private messaging and information sharing – much like the roots, fungi and rhizomes which interconnect surface species in ecological terms (Wohlleben 2016). It is the conditions that austerity has produced that have caused new "social movement rhizomes" (Milan 2018, 162), or rhizomatic commons, to emerge, in a responsive and emotive form, leading community members to reassess not just their current actions, but also to re-evaluate the tenets upon which they had previously grounded their lives. The degrowth perspective can thus be enriched through a contribution which shows that regrowth, in terms of "redefining life's purpose" (Carlsson and Manning 2010, 934), or re-evaluating what makes a useful or fruitful life can emerge at any stage of a person's life. A heterotopian framing enables us to appreciate that these localised responses to austerity politics are orientated around reclaiming power in real, tangible spaces. Unlike utopian visions which are fantastical, stylised and anticipatory, the heterotopian praxis of enacting, doing and making is visible and contemporary. Agency and forms of power are captured through these acts of everyday civic intervention; providing robust examples for others that those considered redundant in contemporary societies – the old, the rural, the unwaged – can instead be highly impactful. Nowtopian practices draw attention to heterotopian spaces. These "different" spaces exist and whereas Foucault's essay considers the othering spaces within cities, the nowtopian examples in this study are exclusively rural. As Johnson notes: "With different degrees of relational intensity, heterotopias glitter and clash in their incongruous variety, illuminating a passage for our imagination" (2006, 87).

What seems to elude current depictions of nowtopian practice is a reflection on the ways in which getting involved in social and ecological non-waged labour might develop from non-political (or non-politically explicit) motives. We need to consider the ways in which non-altruistic intentions or actions that might have originally prompted forms of non-waged civic activities, have consequently stimulated a nowtopian "enlightenment." "Moral work," such as volunteering, caring and vocational support, can cause actors to question the socio-political contexts which require non-remunerative labour (Arai 2000). This implies that the act of working with others to create something which sits outside of capitalist values might in turn be the driver which enables a degrowth perspective to flourish. Indeed, we might go further and propose that such acts of "moral labour" may offer a path to political regrowth. Those who have lived relatively successfully within the conventional economic world can still, once their need for waged income has passed, practice civic acts that are by their nature nowtopian, but not perhaps as overtly political as the nowtopian literature might propose. And yet, in so doing these people are effectively repurposing their understanding and use of what might be referred to as their leisure (and leisure time). On the basis that this apolitical regrowth approach to nowtopianism has yet to gain traction in the literature, the second part of this chapter will build upon empirical research to address to what extent civic voluntarism can offer the opportunity for people to practice a form of heterotopian political regrowth which focuses centrally on the act, rather than outcome, of moral work.

Embedded agency and pragmatic resiliencies in rural communities

Central to the UK government's focus on localism in an era of austerity is the performance of the "resilient" subject. For resilience to flourish subjects must exhibit some form of agency which enables them to champion their own individual capacities and to connect with others. Cinderby et al. (2016) utilise Callaghan and Colton's 2008 work to draw upon their "pyramid of community capital and resilience" (2016, 1256) to explore how six key community assets (economic, built, cultural, social, natural and human) interconnect in dynamic interplay to enrich individual and collective lives. Although Cinderby et al.'s work provides a fascinating overview of a Participatory Action Research community project's life cycle (see McIntyre 2007, for an overview of this methodology), its analysis fails to engage with detailing the characteristics of a resilient subject. Instead we are signposted towards the importance of "community salesmen" or "mavens" (Cinderby et al. 2015, 1267) rather than provided with an explanation of who was involved with their project and their reasons for participation.

The literature then reveals that more fieldwork is necessary to identify resilient subjects and their reasons for action, particularly in a rural context. Within the UK over the last twenty years responses to rural flooding has initiated a growing body of research regarding community resilience. Thus, Sarah Whatmore's study has explored the nuances of community-level knowledge and expertise around local resources (Whatmore 2009; Lane et al. 2011) to interrogate the ways in which local co-produced solutions to flooding informs our understandings of the wealth of vernacular knowledge that exists around water resources. McEwan's et al.'s 2014 work considers the importance landscape and other "expressions of place" have with regards to formations of the self and articulations of individual agency that arise. These forms of vernacular knowledge can be thought of in terms of what Wilson has described as "encoded" learning (2015, 237), whereby specific knowledge related to a particular environment is shared within a community, leading to embedded agency within those community members. How this agency adapts to change seems to be a key facet of resilience.

However, the idea of capacities for civic action residing in rural places is challenged by the stark realities of population change. An issue faced in many rural places is that of ageing, specifically the increase of older people living in rural communities (Scharf, Walsh, and O'Shea 2016) and demographic changes predicted to be experienced in Europe as a result of the rate of youth out-migration overtaking that of in-migration by older groups (Burholt and Dobbs 2012). Whilst this could be viewed as problematic for local governance and service provisioning many argue that these actors contribute greatly to their local communities. The International Rural Ageing Project (1999), an expert review of rural ageing, found that older citizens were a potential untapped civic resource. Their contributions to effective rural policy and planning are argued to be often overlooked in favour of a view of the rural elderly as service consumers instead of active citizens (see also, Munoz et al. 2014). There is then a need to reclaim the term gerontology (the process of ageing) from something negative to a positive embracing of human life stages. Ageing is not necessarily a decrease in social agency, just a different kind of agency. However, there is a continued absence of detailed research on the participation of older people in civic engagement activities and political policy-making (Burholt and Dobbs 2012), making any systematic assessment of the involvement of the elderly in community resilience difficult to assess.

Scharf, Walsh, and O'Shea (2016) have gone some way to address this knowledge gap through providing a detailed synopsis of new developments in current rural gerontology research which they attribute to three factors: a renaissance of "environmental gerontology" (2016, 51) within social gerontocracy, the recognised impacts of globalisation and, lastly, the continued disparity in rural livelihoods compared with urban communities. Their evidence suggests that there is a growing recognition that the rural is not a static place in terms of elder populations. Roberts and Townsend (2016) highlight the adaptive capacities of rural communities. Yet still little attention has been paid to the concept of the types of political power older people influence in rural spaces.

An ageing and expanding rural demographic, along with a recognition of elder civic engagement in political and civic processes, makes it possible to argue that much rural activism concerning the environment is led by cohorts of community elders. These can be termed rural gerontocracies. Given the push within a Localism Act (HM Government 2011) framing for communities to define and enact their own local resilience strategies, coupled with an ageing rural population within the UK, it is not unreasonable to surmise that the task of governing local rural spaces is increasingly led by community elders. Moreover, these elders are asserting their agency through vehicles such as parish councils, flood action groups and environmental campaigns to articulate their disquiet concerning current political and economic processes. The next section of the chapter outlines recent fieldwork undertaken, exploring this in more detail.

The fieldwork

The River Adur catchment was selected for the study, which explored local water resources governance in response to changing environmental conditions. Located in the county of West Sussex, adjacent to the South Downs National Park in South East England, it is mainly rural though only fifty or so kilometres from London. The area is populated with small villages and larger market towns with most of the local economy supported by farming, small businesses and tourism generated by visitors to the National Park. The catchment's villages differ from the local towns as they have a higher number of retired residents, aged 60-plus, encompassing a diversity of socio-demographics, with a mixed housing stock of both large detached homes and smaller former social housing properties. Many of the respondents interviewed had retired to the villages because of the beautiful countryside.

The study site, just within the tidal stretch of the river midway along the catchment, is comprised of three closely located waterside villages: Steyning, Upper Beeding and Bramber. Survey participants were asked to share their experiences of changing water environments within their villages, and their responses or actions to these changes. These participants were recruited on the basis that they lived or worked in the villages and had some interest, understanding or role in local water resources management. Consequently emergency services providers, business owners, parish councillors, farmers, community volunteers, planning officers, householders, property developers, writers and historians were amongst the cohort of participants who took part in the one to one semi-structured interviews. No level of expertise in water resources management was required, simply an interest in local water matters past, present and future. Respondents were asked to talk about their local water environments, leading to an open, generative interviewing format, lasting an hour on average. In all thirty-seven interviews were conducted over the course of the research.

Interviews were transcribed and subject to thematic analysis. First order analysis categorised the data into groupings such as "flooding," "drainage," "pollution," "access," "management" and "governance." However, it became clear that what was manifesting through these interviews was more akin to an oral history, as interviewees' own personal journeys were combined with their interest in water management, local history or environmental governance. As the majority of interviewees were retired, and even for those who were not, asking them to consider local water resources led to a reflection on their own life's path. Time and again participants questioned why the public sector services they had come to expect after years of working and paying tax were no longer assured, leading them to question what would happen to them as they aged – given their sense of a lack of community around them.

Creating a meaningful life in a post-work world

Different interpretations of Foucault's work have suggested that life-stage experiences are an inimical aspect of a heterotopian "othering" space. The counter spaces of the children's playground, the retirement home, the bath house, all are bound by the micro-cultures and particular normativities that operate only in these settings for groups of similarly aged actors. It is argued in this chapter that the phenomenological experience of the "post work" world catapults actors into new life stage temporalities, whereby an almost invisible, incremental series of changes are undergone by those who newly inhabit the realm of the "retiree." Foucault's interest in the term "emplacement" (Johnson 2013, 794) captures this concept of place, occupation (in both sense of the word) and time in explorations of heterotopias. This is particularly pertinent when considering retirement and the physical relocation that often occurs at this life stage. Quite a number of the participants had moved to the area because of the river, nestled as it is within the landscape of the South Downs. Of these, most are retired people for whom the riverbank is also a place of the imagination, where they can contemplate a different life after working life is completed. Other incomers were younger, commuting to work away from the villages. They also valued the river, but perhaps more as a signifier of the "rural idyll" (Mingay 2017), reflecting a better quality of life for them and their families. New relocators, the retirees, often used volunteering in various water resources management capacities as a route to enjoy meaningful activities and build social networks. Many enjoyed a creative resurgence inspired by the surrounding riverine landscape through creating local artwork: visual, oral and aural. Some joined existing local environmental groups, meeting several times a week, either to get involved in the physically hard labour of cutting riverside vegetation or tree planting, and through seasonal citizen observation activities such as bird surveys and river water quality measuring. A large scale one-off activity was in response to degraded local river water quality. In 2015 local anglers and the Ouse and Adur Rivers Trust (OART.co.uk) fund-raised to hire heavy machinery to clear riverbeds and deposit aggregate to help water decontamination and support breeding grounds for spawning fish. Reflecting on the work, several respondents alluded to the need to take direct action, even aligning with those whose perspectives radically differed from their own – because they shared the same ultimate goal, as one of them observed:

> If you get a mobilised, engaged, enthusiastic, driven (Parish Council) chair, the world is your oyster, then you get people that have maybe been doing it for too long, but you just can't find anybody else to take it on . . . so that community engagement thing again is really interesting, and I reflect back into my own life. I have nothing to do with local politics, but I think I have quite a lot to do with my community.
>
> (Local community volunteer co-ordinator,
> male, working non-resident, late 40s)

Other examples of heterotopian activities include volunteers in one village who spent time researching local drainage maps to identify sources of pollution and who held local meetings to attract support and publicity in order to change local land management practices which, they concluded, were the drivers of the problem. Others volunteered as parish councillors, getting involved in civic administration, often of a very technical nature, particularly around the devolution of riparian drainage responsibilities from the regulatory authority (the Environment Agency) to the local homeowners. Their work enabled them to transition out of high status paid jobs and towards useful, but still high status, voluntary work. As one participant observed: "We are ending up running a kingdom almost here." Another group of newly relocated retirees enjoyed a creative relationship with the riverbank by joining walking groups or local art initiatives, or by undertaking solitary endeavours such as writing and thinking using the river as their inspiration.

Talking to these newly retired relocators it was resonant that their move was connected to a determination to shape a meaningful life post-work. Yet more than this was also a wistfulness – of reflecting over a life past, engaged in the headlong rush of work, family and caring – to now re-evaluate what this next stage of their lives might mean for them. Time became the underlying theme behind the research conversations which unveiled the changing focus from working towards long-term goals to appreciating that every day is to be cherished. Many respondents emphasised the joy of finally engaging in their passion for nature conservation, in slowing down time that they felt was passing too quickly. Thus, for instance, one of them explained: "I mean, I just do this for, for love, you know. There's no money, no allowances, there's nothing" (Male, retired environmental conservation volunteer, early 70s). Yet not all of the voluntary activities were supporting health and wellbeing in their myriad forms. For some, volunteering was a necessary response to failing local authorities and a withdrawal of state support. Indeed, as discussed in the next section, volunteering was identified by some interviewees as the only way to protect local habitats.

Resilience in counterpoint to austerity localism

Many residents told stories about the ways in which their lives had been altered by the impact of flood events. In one case the residents of a country lane, far away from the river but nestled within the chalk downland, explained how over the past five winters their homes had become deluged with water sluicing down the tarmac road from the South Downs to pool at the bottom of the lane, washing onto their garden walls and into their houses. As a result, postal and delivery workers would no longer service the street, the residents' cars were damaged by stones and other debris in the water thrown up by passing traffic, and garden walls and brickwork mortar were showing signs of erosion from water intrusion. Although a downland spring had always trickled down the side of the lane, and was a renowned resource for village children to float home-made paper boats along it, the volume of water had now increased substantially in autumn and winter. A collective of activists was formed to keep pressure on the local council. Comprised of six active members, representing eighty local residents, who would meet regularly to plan activities, there were another twelve "sleeper" households who would get involved when they could. The activists collected and submitted photos and engineers' reports showing water billowing along the road, scouring the road surface and making transit difficult for the forty or so households. They identified a combination of changing land management practices, in terms of crop type and tillage, by differing farming practices at the top of the lane, and a withdrawal of local government services leading to the County Council no longer maintaining the highway or the drainage systems in the way that they had been historically. Further, the environmental regulator, the Environment Agency, had also actively withdrawn from managing the drainage maintenance of minor watercourses.

As a result the residents perceived that upland soil was eroding, causing it to cascade from the top of the inclined lane. "Grips" and other small diversion channels that would have traditionally slowed some of the water's progress from the lane and into roadside copses and dips were now no longer being maintained by the municipal Highways Agency. With nowhere to go but straight down, the increased water volume and velocity resulted in a tumult for the residents to deal with. With great candour the residents outlined that the discomfort and inconvenience they suffered should not be a "postcode lottery" as they paid their local taxes for highways services and felt they were being unduly discriminated against:

> We do get quite large chunks of tarmac and stones and things that are washed down . . . when it does rain hard and the water gushes down it brings all the stones down as well . . . there's no real champion if you like from the Council who, who really they say they want to help us out but it's only for their political needs – there's no money around.
>
> (Local resident, newcomer, working, male, early 60s)

For these elders, banding together to both clear the drains of leaves when and where they can, attending and petitioning at parish council meetings, fundraising for and commissioning engineers' reports is all part of a grassroots, responsive endeavour to protect their local environment. The elders are adamant that their voices will be heard, but wistful about how long, as campaigners of retirement age, they can keep their activities going. As one of them pointed out: "Because obviously as people get older they're not, they're less able to, to do these things, aren't they? So you know, that's fine for the next five years but I think after that we'll probably be struggling" (Local resident, late 50s, female, retired).

This connectivity between austerity localism and the more nuanced impacts it has on older people's lives, is resonant also within another flooding incident. Here, flooding came through groundwater discharge, which is common in river valleys with high water tables, leading to the groundwater sitting within saturated soil, close to the land surface. In this instance the water flooded private houses and adjacent business premises. Although rainfall contributes greatly to the amount of water in the soil, the owners felt that the water table had been directly impacted by the rapid and dense housebuilding schemes higher up the catchment. Interviewees – including the householders – cited more roads, more buildings and more hard-standing such as off-road parking, as responsible for the rise in the local water table. One of them said:

> The [flooding] problem . . . wasn't actually caused by the Adur down here. The water just came down slowly through, through the fields; you couldn't really see it but it was below the surface of the grass if you like. You could actually see it coming off the ditch into our garden, you, you could see the flow of it there.
>
> (Retired business owner, female, mid-60s)

Another one went on to explain that "we were out [having to live elsewhere] for, well it was nine and a half months wasn't it?" The householders talked through how flooding had awakened them to austerity politics. When their home flooded they realised there was no-one to help them, no administrative structure or service provider to enable them to return to their prior circumstances. Instead, they felt abandoned by the state, wondering what their taxes, their insurance and all their hard work had been supporting. Their response was to join their local flood action group and to invest a large proportion of their retirement savings into adding flood protection cladding to their property – a property they felt would no longer be sellable as neighbours' sales had been impacted by flooding. As one of them commented, with respect to the neighbours, "Anyway they're trying to sell it and they've had three sale agreed signs up and all removed and we believe, certainly the first two were to do with insurance and inability to get insurance" (same interviewee: retired business owner, female, mid-60s)

Conclusion: activism and community – the creation of heterotopian gerontocracies

Time and again the interviewees reinforced that their desire to keep their community together superseded the challenges of relying upon an ageing cohort of local activists. From running local libraries, to guerrilla gardening, to keeping hedgerows and roadside verges trimmed, to organising village fetes and funding local youth spaces, the weight of responsibility lay on village elders who continue to be the most visibly present and online responsive sector of the local community. This activism is clearly in response to funding cuts and had generated a local gerontocracy in charge of organising the community and managing scant funds from local government. Not only then are these actors enabling heterotopic spaces which run counter to the atomising effects of globalised neoliberalism, but they are demonstrating their capacity as elders to take charge, efficiently, humanely and collectively, of their local governance apparatus. They create, in effect, heterotopian gerontocracies. Their power disrupting techniques are playful, antagonistic and agile – they use their wit, tenacity and experience to make change happen, using a wide portfolio of methods.

As onerous as the workload appears to be there is a tenaciousness from these respondents who recognise the need to fight for their share of local council monies.

Reflecting upon gerontocratic heterotopias reveals that once freed from social hierarchies the "post work" world has the potential to recast work/labour as a productive, life affirming "nowtopian" activity. For many of those interviewed this is the defining, liberating value in the different environmental activist roles that they inhabit. Environmental activism at this third age is connected to this search, desire and need for conviviality and forms of political and social regrowth. This work is joyous as it implies claiming an identity that was, perhaps, left dormant during the years of waged labour (or maybe was never previously claimed). Instead, theirs is a demand to celebrate and enjoy health, nature, the outdoors and physical contact. Rather than hollow consumption, these new individual and collective endeavours are concerned with making and improving – all helping to create new identities and collaborations that sit outside of monetary and consumerist goals. Against the neoliberal paradigm which depicts elders as unproductive and resource depleting, these activists highlight the hollowness of a consumption orientated modernity at a time in their lives when others seek to define them as vulnerable.

The empirical research offered here suggests that these elder environmental activists are part of a heterotopian vanguard. Holding a mirror to normative presumptions concerning ageing and rural vulnerabilities, these disruptive elders are enacting heterotopias on riverbanks in multiple spaces. They contribute to self-producing nowtopias which are appearing in different sites with different foci. Unlike eco-communities or peripatetic gatherings such as Burning Man or NoWhere, however, there are a plethora of ordinary, pedestrian, unrecognised nowtopian practices burgeoning in unrecognised corners. As this chapter argues, we must be careful to be attendant to them or we risk failing to recognise the unique contribution that these elder heterotopians make, both to their communities and as a cohort from which degrowth researchers can learn. More empirical research is therefore needed on the subject of regrowth, to extend current degrowth scholarship and to find examples of elder nowtopian enclaves in other countries and contexts.

The resourceful and imaginative elders in this research have highlighted a pragmatism that comes with age. Having faced the retreat of the state and with a decline in social networks, they have combined approaches which integrate strategies of social insurance along with positive endeavours to make their immediate landscapes and lifeworlds improved for all civic actors. Imagination and agency in ageing are intimately linked, as these older activists recall the riverbanks of their youth, of their cultural imaginary and of their aspirations of now and in the future. Reflecting on the importance of what makes a good life has led to a range of social, physical, emotional and political encounters with others and their surrounding environments. Connected to this is a spontaneous confirmation of degrowth sensibilities. These activists recognise that without strong, connected and empowered communities our lives will continue to be subject to the vagaries of globalised finance. Without a healthy environment, and our own physical and emotional health, our lives are denuded by a morbid fixation on accumulation at all costs. The heterotopian mindsets of these nowtopian practitioners help us see that alternative sustainable futures together are possible.

Works cited

Arai, Susan. 2000. "Typology of Volunteers for a Changing Sociopolitical Context: The Impact on Social Capital, Citizenship and Civil Society." *Society and Leisure* 23 (2): 327–52.

Arendt, Hannah. 1973. *The Origins of Totalitarianism* (Vol. 348). New York: Schocken Books.

Berneri, Marie. 1950. *Journey Through Utopia*. London: Routledge & Kegan Paul.

Brown, Valerie. 2015. "Utopian Thinking and the Collective Mind: Beyond Transdisciplinarity." *Futures* 65: 209–16.

Burholt, Vanessa, and Christine Dobbs. 2012. "Research on Rural Ageing: Where Have We Got to and Where Are We Going in Europe?" *Journal of Rural Studies* 28 (4): 432–46.

Callaghan, Edith, and John Colton. 2008. "Building Sustainable & Resilient Communities: A Balancing of Community Capital." *Environment, Development and Sustainability* 10 (6): 931–42.

Carlsson, Chris. 2008. *Nowtopia: How Private Programmers, Outlaw Bicyclists and Vacant Lot Gardeners Are Inventing the Future Today!* San Francisco: AK Press.

Carlsson, Chris, and Manning Francesca. 2010. "Nowtopia: Strategic Exodus?" *Antipode* 42 (4): 924–53.

Cinderby, Steve, Gary Haq, Howard Cambridge, and Kate Lock. 2015. "Building Community Resilience: Can Everyone Enjoy a Good Life?" *Local Environment* 21 (10): 1252–70.

Collins, Tom. 2016. "Urban Civic Pride and the New Localism." *Transactions of the Institute of British Geographers* 41 (2): 175–86.

Curry, Nigel, Vanessa Burholt, and Catherine Hagan Hennessy. 2014. "Conceptualising Rural Connectivities in Later Life." In *Countryside Connections: Older People, Community and Place in Rural Britain*, edited by Catherine Hagan Hennessy, Robin Means, and Vanessa Burholt, 31–62. Bristol: Policy Press.

D'Alisa, Giacomo, Federico Demaria, and Giorgos Kallis, eds. 2014. *Degrowth: A Vocabulary for a New Era*. London: Routledge.

Edgar, Andrew. 2006. *Habermas: The Key Concepts*. London: Routledge.

Featherstone, David, Anthony Ince, Danny Mackinnon, Kendra Strauss, and Andrew Cumbers. 2012. "Progressive Localism and the Construction of Political Alternatives." *Transactions of the Institute of British Geographers* 37 (2): 177–82.

Foucault, Michel. 1986. "Of Other Spaces." Translated by Jay Miskowiec. *Diacritics* 16 (1): 22–27.

Gearey, Mary. 2018. "Gerontocracies of Affect: How the 'Geographies of Austerity' Have Reshaped Elder Environmental Radicalism." In *Rights to Nature: Tracing Alternative Political Ecologies Against the Neoliberal Environmental Agenda*, edited by Elia Apostolopoulou and Jose A. Cortes-Vazquez. Cambridge: Cambridge University Press.

Gearey, Mary, and Neil Ravenscroft. 2019. "The Nowtopia of the Riverbank: Elder Environmental Activism." *Environment and Planning E: Nature and Space* 2 (3): 451–64.

Glasgow, Nina, and David Brown. 2012. "Rural Ageing in the United States: Trends and Contexts." *Journal of Rural Studies* 28 (4): 422–31.

Guillemot, Jonathan R. and Debora J. Price. 2017. "Politicisation in Later Life: Experience and Motivations of Older People Participating In a Protest for the First Time." *Contemporary Social Science* 12 (1–2): 52–67.

Hetherington, Kevin. 2002. *The Badlands of Modernity: Heterotopia and Social Ordering*. Abingdon: Routledge.

HM Government. 2011. *Localism Act 2011*. Norwich, UK: Stationery Office.

Ingold, Tim. 2011. *Being Alive: Essays on Movement, Knowledge and Description*. Abingdon: Routledge.

International Rural Ageing Project. 1999. *Shepherdstown Report on Rural Aging*. Center on Aging, West Virginia University.

Jameson, Fredrick. 1991. *Postmodernism, or, The Cultural Logic of Late Capitalism*. Durham: Duke University Press.

———. 1979. "Reification and Utopia in Mass Culture." *Social Text* 1: 130–48.

Johnson, Peter. 2013. "The Geographies of Heterotopia." *Geography Compass* 7 (11): 790–803.

———. 2006. "Unravelling Foucault's 'Different Spaces.'" *History of the Human Sciences* 19 (4): 75–90.

Lane, Stuart, Nick Odoni, Catharina Landström, Sarah Whatmore, Neil Ward, and Susan Bradley. 2011. "Doing Flood Risk Science Differently: An Experiment in Radical Scientific Method." *Transactions of the Institute of British Geographers* 36 (1): 15–36.

Lefevbre, Henri. 1991. *The Production of Space*. Translated by Donald Nicholson-Smith. Oxford: Blackwell.

Mannheim, Karl. 2013. *Ideology and Utopia*. Abingdon: Routledge.

McEwen, Lindsay, Owen Jones, and Ian Robertson. 2014. "'A Glorious Time?' Some Reflections on Flooding in the Somerset Levels." *The Geographical Journal* 180 (4): 326–37.

McIntyre, Alice. 2007. *Participatory Action Research*. Los Angeles, London, New Delhi and Singapore: Sage Publications.

Milan, Sonia. 2018. "Data Activism as the New Frontier of Media Activism." In *Media Activism in the Digital Age*, edited by Victor Pickard and Guobin Yang, 151–64. London and New York: Routledge.

Milbourne, Paul. 2016. "Poverty and Welfare in Rural Places." In *Routledge International Handbook of Rural Studies*, edited by Mark Shucksmith and David Brown, 450–61. London: Routledge.

———. 2015. "Austerity, Welfare Reform, and Older People in Rural Places: Competing Discourses of Voluntarism and Community?" In *Ageing Resource Communities: New Frontiers of Rural Population Change, Community Development and Voluntarism*, edited by Mark Skinner and Neil Hanlon, 96–110. London: Routledge.

———. 2012. "Editorial: Growing Old in Rural Places." *Journal of Rural Studies* 28 (4): 315–17.

Mingay, Gordon. 2017. *The Rural Idyll*. London: Routledge.

Moore, Jason. 2017. "The Capitalocene, Part I: On the Nature and Origins of Our Ecological Crisis. *The Journal of Peasant Studies* 44 (3): 594–630.

Munoz, Sarah-Anne, Jane Farmer, Jeni Warburton, and Jenny Hall. 2014. "Involving Rural Older People in Service Co-production: Is There an Untapped Pool of Potential Participants?" *Journal of Rural Studies* 34: 212–22.

O'Hara, Mary. 2015. *Austerity Bites: A Journey to the Sharp End of Cuts in the UK*. Bristol: Policy Press.

Proshansky, Harold. 1978. "The City and Self-identity." *Environment and Behaviour* 10 (2): 147–69.

Roberts, Elisabeth and Leanne Townsend. 2016. "The Contribution of the Creative Economy to the Resilience of Rural Communities: Exploring Cultural and Digital Capital." *Sociologia Ruralis* 56 (2): 197–219.

Scharf, Thomas, Kieran Walsh, and Eamon O'Shea. 2016. "Ageing in Rural Places." In *International Handbook of Rural Studies*, edited by Mark Shucksmith and David Brown, 50–59. Abingdon: Routledge.

Schumacher, Ernst Friedrich, and Peter N. Gillingham. 1979. *Good Work*. New York: Harper & Row.

Soja, Edward. 1995. "Heterotopologies: A Remembrance of Other Spaces in the Citadel-LA." *Postmodern Cities and Spaces*: 13–34.

Stockdale, Aileen, and Marsaili MacLeod. 2013. "Pre-retirement Age Migration to Remote Rural Areas." *Journal of Rural Studies* 32: 80–92.

Stokowski, Patricia. 2002. "Languages of Place and Discourses of Power: Constructing New Senses of Place." *Journal of Leisure Research* 34 (4): 368–82.

Tuan, Yuan. 1977. *Space and Place: The Perspective of Experience*. Minnesota: University of Minnesota Press.

Walker, Alan. 2010. "The Emergence and Application of Active Aging in Europe." In *Soziale lebenslaufpolitik*, 585–601. Wiesbaden: VS Verlag für Sozialwissenschaften.

Whatmore, Sarah. 2009. "Mapping Knowledge Controversies: Science, Democracy and the Redistribution of Expertise." *Progress in Human Geography* 33 (5): 587–98.

Wilson, Geoffrey A. 2015. "Community Resilience and Social Memory." *Environmental Values* 24 (2): 227–257.

Wohlleben, Peter. 2016. *The Hidden Life of Trees: What They Feel, How They Communicate – Discoveries from a Secret World*. Minneapolis: Greystone Books.

5 Agricultural heterotopia

The Soybean Republic(s) of South America

Gladys Pierpauli and Mariano Turzi

Introduction

This paper contends that the current model of international agriculture – global agribusiness – has pressed towards consolidating an integrated, export-oriented productive entity throughout South America: the "Soybean Republic." The economically driven transformation of geographical domains is the necessary result of their incorporation into the broader economic space of globalisation. At the same time, the homogenising tendencies of global, liberal, and capitalist agriculture are simultaneously represented, contested, and inverted in the "Soybean Republic," which we argue constitutes an "agro-heterotopia." In analysing the structure and dynamics of this heterotopia in this chapter, we focus particularly on the Brazil-Paraguay soybean frontier. In so doing, we reveal how the differences and disjunctures produced by seemingly homogenising globalisation processes bear upon global agricultural capitalism. The discursive and physical space of the Soybean Republics is transected by neoliberal flows and discourses connected to the modern and mechanised global order of capitalist agribusiness; at the same time, the Soybean Republics are contested by local, ancient, indigenous, and peasant resistance, which refuses this model of agricultural production.

In putting forward the term Soybean Republic, we take our inspiration from a 2001 advertisement for a plague monitoring system from the Swiss chemical company Syngenta (Turzi 2017). "Soybean knows no boundaries," the advert declared, presenting a map in which a green splotch is superimposed over parts of Argentina, Brazil, Paraguay, and Bolivia. To conceptualise this space, we propose an alternative name for this entity: the "Soybean Republics." This shift to the plural is meant to set up a contrast between agricultural utopia and heterotopia in the context of twenty-first century globalisation. Official agricultural reports produced by international organisations carry notices such as the following, from the United Nations:

> Disclaimer: The designations employed and the presentation of the material in this publication do not imply the expression of any opinion whatsoever on the part of the United Nations Environment Programme concerning the legal

status of any country, territory, city or area or of its authorities, or concerning
delimitation of its frontiers or boundaries.

This disclaimer can be construed as expressing a utopian impulse on the part of
international institutions, which support the construction of a smooth and "uninterrupted" – indeed, perhaps uninterruptible – form of globalisation. In the specific case of agriculture, this means imposing homogenising agricultural policies
worldwide in the belief that such a "cookie cutter approach" can be implemented
without regard to geographical, political, or territorial specificities. Here, utopia
is what Foucault described as a "placeless place" (Foucault 2000 [1967], 4). In
reshaping territorial realities according to the model of international agribusiness,
however, this utopia was actually emplaced, and has thus become a heterotopia –
what Foucault called "an effectively enacted utopia" (Foucault 2000, 3).

This chapter is divided into two sections. The first analyses the Soybean
Republic as a utopia in that it involves the spatial and discursive implantation of
a political and economic global order. This section explains how and why global
agribusiness became the dominant model of agricultural production and how it
reshaped geographical realities according to its own neoliberal imperatives of
ensuring comparative advantage and integration into the world economy. The
second section introduces the Soybean Republics' critical heterotopian reverse
side. In approaching the Soybean Republics as both a utopia and heterotopia,
we do not mean to present these two concepts as inherently opposed, but rather
at variance or in contradiction with one another. Discourse and space overlay at
the South American soybean frontier, constructing divergent domains. Foucault
describes heterotopia as "utterly" different "from all the emplacements that they
reflect or refer to" (Foucault 2000, 2). Still, in their difference, heterotopias are not
as static sealed totalities (Saldanha 2008), but rather sites of dynamic, diacritic,
ever changing, and conflictive processes.

Our deliberate contrast between the concepts of utopia and heterotopia highlights the juxtaposition of "incompatible emplacements" that defines heterotopic
space. The proposed conceptual interplay seeks to highlight the relational dimension of heterotopias. The essence of the relation between utopia and heterotopia
lies in this reciprocal play of tensions. The Soybean Republics do not sit in isolation as reservoirs of freedom, emancipation, or resistance; nor is the Soybean
Republic sealed off as a repressive utopia. The complex relations at work in our
case study encapsulate the ways in which, for Foucault, heterotopias are "connected with all other emplacements, but in such a way as to suspend, neutralize, or
invert the set of relations designated, reflected, or represented by them" (Foucault
2000, 2). Despite exhibiting their own separate, distinct logics and categories, the
Soybean Republic and the Soybean Republics are not irreconcilable opposites.
These two articulations of South American agriculture combine, and connect. The
relations between economy, community, and ecology diverge and coexist in the
same polity. Social and institutional arrangements overlap, producing a state of
permanent tension. Technologies of space and time are in constant dialogue and
disagreement in the process of spatial ordering. As heterotopias, the meaning of

the Soybean Republic and Soybean Republics meaning cannot be grasped separately. Nor are they stable entities; rather, they are contingent, provisional, and evolving.

Utopia: the Soybean Republic

The increasing internationalisation of production processes has been among the most important transformations in the world economy since the early 1970s. Progressively, the way of producing goods and services has been modified, such that they are now predominantly organised in what are known as the Global Value Chains (GVCs). GVCs are networks of labour and production processes which result in finished commodities (Hopkins and Wallerstein 1994). The emergent patterns of geographic and economic restructuring and governance that result from this shift to GVCs require that each stage or task in production is situated in a different global location, depending on the availability, quality, and price of the necessary resources and skills. Different stages in the production process, therefore, take place in different economies. Intermediate inputs, such as parts and components, are produced in one country and then exported to others for further production and/or assembly (Gereffi, Humphrey, and Sturgeon 2005; Pietrobelli and Rabellotti 2007). The sum of activities involved in the production of a given commodity constitutes the product's chain or complex (Gereffi and Korzeniewicz 1994).

As a result, the last four decades of international industrial organisation have been characterised by the increasing geographical fragmentation of production, with a few lead firms and traders managing functional integration. In this new global division of labour, lead firms coordinate the activities of upstream and downstream business partners. These developments reorganised spaces according to a corporate-driven model of organisation of production and space. Alongside the economic benefits (for a minority), there have also been overwhelming costs: environmental degradation, poor working conditions, poverty, inequality, lack of access to public services, discrimination, or even national security concerns.

Agriculture was not immune to this transformation, undergoing its own transition to GVCs in the 1990s. Multinational chemical and biotech companies precipitated this dramatic change by adopting the "soybean package" (which combined genetically modified seeds with agrochemicals and direct sowing). In this system, more efficient and better-financed producers, who have thus been able to adopt the soybean package, enjoy cost advantages. This has led to the progressive adoption of the soybean package as the standard of global agribusiness production. The concept of agribusiness can be traced back to John H Davis (Davis 1956; Davis and Goldberg 1957). Agribusiness includes the sum of all operations involved in the manufacture and distribution of farm supplies: production on the farm, storage, processing, and distribution of farm commodities and derived products. Agribusiness liberalises agriculture: food becomes integrated into the economic and productive system. Liberalised markets and scale advantages generated opportunities and incentives for the vertical integration of the soybean trade, from farm to fridge. The soybean chain is a vertebra in a larger grain GVC, which is itself a link in a larger agrifood GVC. As agriculture has globalised, concentrated corporate multinational actors have come to dominate food production, processing, and distribution. According to 2018 research from the University of California, Berkeley, three major corporate mergers concentrated control over international markets for agricultural chemicals, seeds, and fertilisers in three massive corporate actors: Bayer-Monsanto, DuPont-Dow Chemical, and ChemChina-Syngenta. Drawing on their financial strength, such corporate entities have even been able to dictate infrastructural developments in the soybean producing areas of South America. In this way, the implantation of this global model has rearranged geographical space according to market imperatives.

We claim that GVCs are the *global utopia* of twenty-first-century capitalism. Foucault's conception of utopia is predominantly centred on what could be characterised as an unreal space that reflects either a perfectly ordered society or the opposite of society. Foucault (2000, 3) defines utopias as "emplacements that maintain a general relation of direct or inverse analogy with the real space of society." Agribusiness in South America is driven by a neoliberal utopian vision of a capitalist Soybean Republic. In this utopian vision, agribusiness reportedly combines an agronomic perspective ("feeding a growing population"), environmental perspective ("saving the planet"), sociological perspective ("sustainable rural livelihoods and social equity"), and an economic perspective ("efficiency, productivity, and cost effectiveness"). This in itself has deep historical roots that go all the way back to colonial times and the seizure of Latin America as the "New World." The utopian condition – the idea of a "placeless place" – is still operative in the twenty-first century in the form of globalising neoliberal economic paradigms. Accordingly, the Soybean Republic should be seen as being much more than an economic, GVC-driven emplaced production process; it is a totalising social vision.

Public discourse and institutional practice in the Soybean Republic do not fundamentally challenge either this utopian vision or its technological and economic basis. The utopian discourse articulated by agribusiness frames issues such as environmental sustainability narrowly, only mentioning them insofar as they pertain to production. Agribusiness leaders express pride that, thanks to the adoption of the "soybean package," the country contributes to global food production. The harmful effects on society and the environment are sidelined or downplayed. Social issues – poverty, inequality, and labour rights and standards – are repressed in official discourses, while alternative values such as family agriculture are marginalised. The utopian vision implicit in global agribusiness frames agricultural practice in binary terms: environment versus economics; tree huggers versus workers wielding chainsaws; conservationists versus profit seekers; the weak, green, and idealistic versus the strong and pragmatic *fazendeiros* (landowners).

The teleological march of the Soybean Republic utopia is registered in the changes that it has brought about in both environmental management and economics. Contrary to the image of nature as requiring protection prevalent among environmental organisations, government officials of soybean-producing states frame the forest as something backward and brutish, an entity to battle against and exploit. In a 2009 interview, Rui Prado, then President of the Mato Grosso Farm Bureau (FAMATO), articulated what he called an "economic ecology": "To condemn a whole region to backwardness is not fair," he claimed. "No production means no energy, no roads, no railroads, and no ports. Without the state we would have no taxes, but also no infrastructure, or governance." On the same occasion Blairo Maggi, former Governor of Mato Grosso and later national Minister of Agriculture, declared that a farmer in the red cannot take care of the green. The jungle is not better than the city; the jungle means backwardness."[1]

State policy in the Amazon, then, has followed the modernist mindset epitomised in the national motto inscribed on the Brazilian flag: "Order and progress," achieved through the sovereign state and economic development. The transformation of the Amazon's formerly unexploited territory into a "civilised" space implies turning "natural" areas into "developed" ones through the political assertion of territory and economic reshaping of space. In this way, development policy discourses inscribe on the Amazonian a narrative that progresses from lawlessness to order and normality, from undeveloped to developed, backward to modern, poor to rich. The statements given earlier indicate how Brazilian agribusiness leaders see themselves as "self-sovereign" (Moore 2005), that is, assertive subjects endowed with the material means to impose form and meaning on environments. The multiple ways in which geography is concretely experienced – including all of its meanings in ancient, indigenous, and peasant cultures – is obliterated in favour of an efficiency-driven marketisation of agricultural practices. The homogenising power of agribusiness transforms heterogeneous *place* into flat, uniform, and homogeneous *space*.

The socioeconomic transformation of agriculture resulted not just from state policies alone, but from the interaction among policymakers, local elites, and the military. Far from being locally bounded, this occurred through transnational flows, actors, and connections. In the utopia projected by agribusiness, a rolling frontier moves across the landscape, continually transforming space from forest to cropland. Kleinpenning (1987) has analysed how, as part of this process of agricultural development, the Paraguayan government promoted and facilitated Brazilian immigration to Paraguay. This process of economic colonisation was based not on individuals' economic decisions. Rather, it marked by the encroachment of Brazil, an ostensibly superior and more developed country, upon Paraguay, in the same way that supposedly "better" European countries dominated Latin America.

As would be expected in a heterotopia, colonisation creates incongruity. New populations settle in distant areas with which they are unfamiliar and which they imagine are uninhabited. Discourses promoting agricultural modernisation seek to justify the occupation of supposedly virgin territory by presenting agribusiness as the agent of economic progress and political governance, and equating peasant agriculture or indigenous peoples' ways of life as brutish and backward. Ignoring the traditional forms of legitimacy claimed by preexisting native groups in this way implies the foreigner's superiority. Among the *brasiguayos*, a common utopian vision looms into view: they come from a more civilised and modernised country, and possess a more advanced knowledge of agriculture. Through this narrative of superior development, these landowners legitimise themselves. Once dominated by European colonisers itself, Brazil has now come to dominate and colonise Paraguayan land. Wealthy, foreign agricultural landowners are confiscating the land of small-scale Paraguayan farmers, peasants, and indigenous peoples.

Foucault emphasises how utopias are "society perfected or the reverse of society"; as such, they "are fundamentally and essentially unreal" (Foucault 2000, 3). If the Soybean Republic is "essentially unreal" then it cannot be achieved by definition. Still, the utopian vision of the Soybean Republic drives and shapes development in South America. As this utopia is realised and implemented in practice – as it is emplaced – it becomes a heterotopia. In this process of realisation, it fragments and adapts to specific local conditions. As such, it is best described as the "Soybean Republics."

Heterotopia: the Soybean Republics

The previous section established how the embedding of an international, corporate-driven model of organising agricultural production has reshaped territorial realities according to global production demands (Lefebvre 1991). Modes of production evolve from the contradiction between the means (material forces of production) and (social) relations of production (Harris 1989). In this tension, legal, ideological, and institutional structures attempt to realise certain distributions of power in society (Foster-Carter 1978). According to Marxist theory, a mode of production encompasses the totality of the social and technical human interconnections involved in the social production and reproduction of material life (Bottomore 1991).

In the interstices between material forces, such as the international biotechnological revolution, corporate land concentration, and transnational capital, and social relations, such as the local unevenness of the social and political space, the utopian Soybean Republic fragments into the heterotopian Soybean Republics. Foucault's account underscores how heterotopias are not static, but emerge through a play of relations and resemblances. Agro-heterotopias are constituted by meanings layered onto a single material site which juxtapose multiple, even incompatible places with a single space. The distorted mirror image of the Soybean Republic emerges in a heterotopian form: small-scale, family peasant agriculture clashing with big, corporate agribusiness MNC's; mass food production superseding local supply; foreign immigrants as agents of economic wealth in tension with local populations condemned to poverty; neoliberal international success against national failure; and a seamless flow towards progress through globalisation versus the static stagnation of locally rooted models. One of the most telling examples is the battle over seeds. Genetically Modified crops are promoted as a replacement for all other local varieties, which enables large companies to control seed supply and charge royalties on intellectual property. The homogenising drive that is challenged and contested in heterotopian fashion seeds are illegally copied and adapted to local conditions.

As such, the Soybean Republics reflect the heterogeneous quality of this region, the discontinuity and changeability hidden behind the stable and bounded vision that underpins the Soybean Republic. Indeed, by pluralising this Republic, we stress that these agro-heterotopias reflect the inherent tensions embedded in the social structures, ideas, interests, and institutions surrounding agriculture, markets, and power in South America. As heterotopias, the Soybean Republics appear to simultaneously "represent, contest, and invert" global agriculture (Foucault 2000, 3). Implicitly, heterotopias enact a critical commentary on global agriculture and, by extension, globalisation as such.

As heterotopian "counter-sites" (Foucault 2000, 3) that both reflect and disrupt surrounding social space, the Soybean Republics exist in a perpetual state of tension, primarily over land tenure in Paraguay. These disputes arose as a result of Alfredo Stroessner's long dictatorship (1954–1989). Stroessner consolidated personal authority over the institutions of the state and Paraguayan political system through a functional division of labour. The Colorado party, to which he belonged, ran the government and bureaucracy, while the military repressed any political opposition or dissent. A system of cronyism and corruption (clientelism and bribes) interfered with agricultural practice. In rural areas, the regime never perceived peasants or indigenous populations as a political or military threat. With the vast majority of the rural population speaking Guaraní and living from subsistence farming, they were politically invisible.

The expansion of the Paraguayan economy in the 1960s and 1970s transformed environments and ecologies into economic assets. In this expansion of the agricultural frontier, the Chaco region was developed into cattle ranches and cotton fields. The government launched a program to increase production, relieve population pressure, and encourage agricultural modernisation (Nickson 1981). In 1963, the Institute of Rural Welfare (IBR) was created to organise this process. Its main task was to remove – the officially accepted word was "relocate" – squatters and poor farmers from new agricultural colonies in the north and eastern regions. This had been a longstanding demand among landowners. Its realisation brought about the forced resettlement of local farmers and peasants, the majority of whom had a precarious legal hold on their lands or were simply squatting. To secure his hold on the country, Stroessner redistributed land in a way that co-opted the political elite and kept the military loyal. A total of 6,744,005 hectares of land were unscrupulously gifted to friends and political allies of the regime (CVJP 2008). As part of this process, between the early 1960s and mid-1980s, the Paraguayan agricultural frontier underwent a process of "intense colonisation" (Neupert 1991) in which space was materially and discursively contested. Almost a million hectares of agricultural property were illegally registered after the fall of the military government in the period 1989–2003. The dictatorship's rationality of spatial ordering in rural areas demanded the repression of "deviant" groups – most of them indigenous and peasant. These groups themselves represented a challenge to the pillars on which the agribusiness model is built: the industrial, large-scale globalised mode of production; land ownership patterns; governance structures; and cultural practices.

Another heterotopian aspect of the Soybean Republics is the migratory diversity of this area, the concurrence of socioeconomic groups in an agricultural heterotopia. The largest population of foreign migrants in the rural areas of Paraguay is Brazilian, although other groups such as Japanese and Mennonites are also well established. The Korean Unification Church bought 240,000 hectares of land, which included an entire municipality with 6,000 inhabitants in the year 2000. A diverse population does not in itself make for a heterotopia. Social, political, and economic diversity is a necessary, but not sufficient condition for heterotopia. Indeed the Soybean Republics reflect and refract broad tensions in the contemporary agricultural world: modern versus traditional, ecological sustainability versus economic gain, concentration versus dispossession, exploitation versus habitation, development versus settlement, and efficiency versus equality. As such, they involve the overlay of incompatible factors in one geographic space: chemically-intensive technology, the economic concentration of assets, large extensions of land combines, and the expulsion of indigenous people and poor Paraguayan peasants.

The state-sanctioned policy of giving *Brasiguayos de facto* territorial and economic control has created fundamentally incompatible patterns of territorial normalisation. Almost the entirety of the border between Brazil and Paraguay belongs to companies and people of Brazilian nationality or descent. Within this fluid border territory, multiple belongings fold into one another. The *brasiguayo* community's demographic predominance, economic importance, and political governance have led to cultural dominance. The agents of agricultural displacement and border occupation, the *brasiguayos* that own much of the soybean chain have turned economic importance into political power. In this way, export-oriented soybean production has created a new class of agricultural barons.

The resulting social stratification has redistributed power in the rural structure. This historical process has reshaped territory in a manner laden with heterotopian tensions among social classes and ethnic groups, all locked in mortal conflicts over land. The indigenous and peasant settlements attempt to create alternative social orders, while local police, paramilitary groups, and sometimes landowners themselves have been responsible for the disappearance, assassination, extortion, forced eviction, and displacement of peasants and indigenous populations. While the Constitution includes a legal, formal recognition of indigenous peoples' right to land, economic reality and political informality undermine legal procedure. The Inter-American Court of Human Rights (IACHR 2006) has compelled the Paraguayan government to address this lack of adequate procedure. Systemic inadequacies in the land registry system prevent a reliable inventory of landholdings. In a glaringly heterotopian display, registered land in Paraguay exceeds the size of the country. Such a discrepancy represents an attempt to create a neoliberal agricultural subjectivity uprooted from geographical reality. In this collision between official record keeping and everyday place-making, socioeconomic relations and practices simultaneously juxtapose and rearrange power relations. Concretely, written law is rendered moot in Paraguay's rural areas. Regulations are tailored to serve landowners, local courts favour the interests of the landed *brasiguayos*, and police act as private security guards rather than public servants.

The space of the state generates the official utopian vision of agribusiness, whereas the public sphere challenges and resists this vision through illegal practices, occupying landowners' estates or squatting on contested lands being two pronounced examples.

According to the Paraguayan Peasant Movement (MCP), the umbrella organisation for peasants, there were more than 450,000 landless peasants in the country in 2019. Moreover, in rural areas, the proportion of workers employed in the informal economy reached 78% of the total rural labour force in 2018 according to United Nations data.[2] Populations are highly precarious, having no access to social or workplace protection, and only partial coverage by labour laws. This makes them vulnerable to exploitation, including contemporary forms of slavery, and serves to exclude much of the Paraguayan population from the agribusiness utopia. Peasant organisations challenge the Soybean Republic utopian discourse that legitimates agribusiness. For example, they reject the modern-infused term "agribusiness" and replace it with the label "extractivist agricultural model" (Burchard and Dietz 2014). This difference in nomenclature attempts to stress environmental and social cost rather than economic gain. The notion that the utopia of the Soybean Republic is the only possible model of agricultural production is challenged by local peasants and indigenous groups. Rural resistance organisations contest the export-oriented, foreign-owned, and chemically-intensive extractivist monoculture of agro-industry. They advocate protecting traditional, family-based smallholder agriculture, which sustains local livelihoods and protects biodiversity.

Utopian agribusiness discourse, in contrast, stresses the benefits of market-driven agriculture for productivity and growth, arguing that this feeds into national economic wellbeing in the form of abundant food. Competitive food export prices are promoted in this hope that they might lead to current account surpluses through a rise in export values. Moreover, they are endorsed by public policy, on the grounds that they boost gross national income. The agribusiness sector is thus presented as a model of successful entrepreneurship, an engine of economic development, and a critical aspect in political governance. Attempts to contest the collusion of state and global actors entailed in the Soybean Republic utopia have been unsuccessful. Paraguayan indigenous or peasant resistance efforts have not been integrated into global strategies of confronting development and establishing environmental sustainability, such as has been the case with the Huaorani indigenous peoples in the Ecuadorean Amazon (Martin 1999).

The Soybean Republics would constitute what Foucault calls "heterotopias of deviation": "Those in which individuals whose behaviour is deviant in relation to the required mean or norm are placed" (Foucault 2000, 4). Indeed, an elementary characteristic of this particular heterotopia is that illegality has become a "functional deviance": illegal activities make possible an ordinary way of life (Otsuki 2012). Illegality – a structural organising principle of socioeconomic activity in the Soybean Republics – has resulted in a politics of state absence, for corruption is prevalent among law enforcers (Hochstetler and Keck 2007, 151). The enduring illegality imputed to the land and labour of the peasantry and indigenous peoples in the Soybean Republics – the spatial concentration of societal deviance – is typically heterotopian. Indeed, illegality of this kind should be seen as what Pendleton calls "functional deviance" in that it constitutes the structure through which economic interests, social identity, and political institutions interact (2007, 19).

Conclusions

Engaging the Soybean Republics of South America as a concrete articulation of heterotopia has revealed that the current phase of globalisation has created a transnational, corporate-dominated mode of agricultural production in South America. Homogenisation is the dominant discursive and spatial manifestation of global agribusiness. Seen through a heterotopian lens, however, a more complex interaction and dynamic overlay of ideas, interests, and institutions appears. Grasping the Soybean Republics as a heterotopia – indeed, a heterotopia of deviation – reveals the multiplicity of sites involved and social relations in play as Paraguay is integrated into the system of global agriculture.

What we have called the *Soybean Republic* signifies a model of global agriculture that has undifferentiated geoeconomic space. It has done this by compromising integral national spaces and identities, undermining sovereign territories and political boundaries. This pattern of organising the geoeconomic space of agriculture does not adhere to political limits or barriers. Soybeans flow in a continuous stream from Paraguay to Argentina, and agricultural machinery comes into Paraguay from Brazil as if national boundaries do not exist. This continuous material flow also encompasses money and individuals. Indeed, this includes a fluid, immaterial identity: the *brasiguayo* self. Despite its far-reaching effects, the utopian vision underpinning the Soybean Republic has proven to be a sterile and decontextualised aspiration, which preserves rather than challenges existing relations of power.

Crucially, however, this global economic force – bolstered by a (neo)liberal narrative of transnational flows – has not annulled received forms of social, geographical, and economic difference. Instead, it has produced a new heterotopia. In the agro-heterotopia of the Soybean Republics, globalisation has preconditioned clashes between incompatible elements that were once distant and disconnected: civilisation versus barbarism, progress versus backwardness, technology versus tradition, local versus global. As such, the Soybean Republics represent an example of the new kinds of heterotopian space produced in the globalised world of the twenty-first century.

The heterotopian form assumed by these South American territories brings us back to the uncomfortable yet unavoidable complexity of spatial and discursive difference. Having first been brought about by globalisation, the heterotopia interrupts globalisation, unsettling its dominant patterns and paradigms. The Soybean Republics question the liberal dominant rearrangement of global space according to dictums such as the "invisible hand" of international economics, global markets, and comparative advantage. This agricultural heterotopia interrupts the seamless material flow and immaterial narrative of globalisation by bringing otherwise latent economic, environmental, social, and political tensions to the fore. Consider how ancestral rights to use the land challenge landowners; the traditional model of family agriculture advanced by peasants challenges agribusiness; and environmentalists seeking to protect forests challenge the expansion of the agricultural frontier.

Indeed, a key part of our analysis has been that the significance of the Soybean Republics emerges from their existence in relationship to places and processes beyond their borders. This uneasy coexistence is an essential way in which heterotopia interrupts globalisation. Even if ultimately unsuccessful, resistance to integration into the broader system of global capital nonetheless becomes manifest in heterotopia. The peasant community of Jejuí in the San Pedro Department is still a symbol of defiance to the Stroessner dictatorship. In 1969, this peasant community pursued the goal of creating a settlement based on community land ownership and associative work, supported by the Church-sponsored Agrarian Leagues. This stood in stark opposition to agribusiness's model of rural development, as promoted by the dictatorship, which prioritised large landowners' estates. The regime was especially concerned that the example of Jejuí could be replicated in other colonies in the area. Thus, on February 8, 1975, agents commanded by the head of the police arrested 600 people, tortured the group's leaders, and razed the houses of the 29 families that inhabited Jejuí. The correlation between economic landed interests, weak institutional structures, and illiberal forms of governance is ubiquitous in Paraguay, even today. In our case study, the differences through which the heterotopias of the Soybean Republics articulate a distinct social order, include tensions among global agribusiness, local peasant/indigenous agriculture, state sovereign power, and native resistance. By interrupting dominant agricultural practices and policies, this agro-heterotopia embodies the punishing costs of the agribusiness model: namely, environmental degradation due to agricultural frontier expansion into the Amazon and soil contamination following the overuse of agrochemicals; MNC's economic domination and local elites' hold on institutions; social exclusion and the disenfranchisement of much of the population (especially indigenous peoples and local peasants); and the legitimation of state violence as a means to secure advancing progress and development. Moreover, the Soybean Republics enact a critical commentary on the current structure and dynamic of globalisation. Contradicting the idea that globalisation represents the seamless triumph of market capital and creates a harmonious commercial world, the Soybean Republics act as a window into globalisation as it (often) really is, evincing differences and disjunctures in globalisation processes.

In this chapter, we have broadened the discourse on heterotopia to encompass the agricultural frontiers of the Global South. The Soybean Republics mirror and at the same time distort, unsettle, and invert the structure and dynamics of contemporary globalising processes. In tracing the contours of a heterotopia in the contemporary rural world, we hope to have illuminated how a series of contradictions articulate together uneasily in contemporary globalisation. In the Soybean Republics, local residents clash with multinational corporate actors; the imperatives of global profit run up against the friction offered by situated realities; while the promises of cosmopolitan liberal rhetoric belie the ground truth of agricultural colonisation.

Notes

1 The four quotes – Weirich, Prado, Maggio, and Arioli Silva – were taken from authors' interviews carried out during field research in Brazil (August 2008–July 2009).
2 Report of the Special Rapporteur on contemporary forms of slavery, including its causes and consequences, on her mission to Paraguay, A/HRC/39/52/Add.1, 20/7/2018.

Works cited

Augé, Marc, and Sara Mackian. 1995. *Non-places: Introduction to an Anthropology of Supermodernity: From Places to Non-places*. London: Verso.

Beckett, Angharad E., Paul Bagguley, and Tom Campbell. 2017. "Foucault, Social Movements and Heterotopic Horizons: Rupturing the Order of Things." *Social Movement Studies* 16 (2): 169–81.

Boddy, Kasia. 2007. "The Modern Beach." *Critical Quarterly* 49 (4): 21–39.

Bottomore, Tom et al. 1991. *The Marxist Thought*. Oxford: Blackwell.

Brenner, Neil, and Nik Theodore. 2002. "Cities and the Geographies of 'Actually Existing Neoliberalism'." *Antipode* 34 (3): 349–79.

Burchardt, Hans-Jürgen, and Kristina Dietz. 2014. "(Neo-) Extractivism – A New Challenge for Development Theory from Latin America." *Third World Quarterly* 35 (3): 468–86.

Cleary, David. 1993. "After the Frontier: Problems with Political Economy in the Modern Brazilian Amazon." *Journal of Latin American Studies* 25 (2): 331–49.

Cleave, Chris. 2009. *The Other Hand*. London: Hachette.

CVJP (Comisión de Verdad y Justicia del Paraguay). 2008. *Informe Final Tierras Mal Habidas*. Paraguay: Tomo IV. Asunción.

Davis, John H. 1956. "From Agriculture to Agribusiness." *Harvard Business Review* 34 (1): 107–115.

Davis, John. H. and Ray A. Goldberg. 1957. *A Concept of Agribusiness*. Boston: Harvard University, School of Business Administration.

Dehaene, Michiel, and Lieven De Cauter, eds. 2008. *Heterotopia and the City: Public Space in a Postcivil Society*. Abingdon: Routledge.

Faramelli, Anthony, David Hancock, and Robert G. White, eds. 2018. *Spaces of Crisis and Critique: Heterotopias Beyond Foucault*. London: Bloomsbury.

Foster-Carter, Aidan. 1978. "The Modes of Production Controversy." *New Left Review* 107 (1): 47–78.

Foucault, Michel. 2012. *Discipline and Punish: The Birth of the Prison*. London: Vintage.

———. 2000 [1967]. "Different Spaces." In *The Essential Works of Foucault 1954–1984: Aesthetics, Method, and Epistemology*, edited by James Faubion. London: Penguin.

Genocchio, Benjamin. 1995. "Discourse, Discontinuity, Difference: The Question of 'Other' Spaces." In *Postmodern Cities and Spaces*, edited by S. Watson and K. Gibson, 35–46. Oxford: Blackwell.

Gereffi, Gary, John Humphrey, and Timothy Sturgeon. 2005. "The Governance of Global Value Chains." *Review of International Political Economy* 12 (1): 78–104.

Gereffi, Gary, and Miguel Korzeniewicz, eds. 1994. *Commodity Chains and Global Capitalism*. London: Praeger.

Harris, Marvin. 1989. *Cows, Pigs, Wars, & Witches: The Riddles of Culture*. New York: Vintage.

Harvey, David. 2007. *A Brief History of Neoliberalism*. Oxford: Oxford University Press.

Hetherington, Kevin. 2002. *The Badlands of Modernity: Heterotopia and Social Ordering*. Abingdon: Routledge.

Hochstetler, Kathryn, and Margaret E. Keck. 2007. *Greening Brazil: Environmental Activism in State and Society*. Durham: Duke University Press.

Hopkins, Terence K., and Immanuel Wallerstein. 1994. "Commodity Chains: Construct and Research." In *Commodity Chains and Global Capitalism*, edited by Gary Gereffi and Miguel Korzeniewicz. London: Praeger.

Hurrell, Andrew. 1991. "The Politics of Amazonian Deforestation." *Journal of Latin American Studies* 23 (1): 197–215.

Inter-American Court of Human Rights. 2006. *Case of the Sawhoyamaxa Indigenous Community v. Paraguay*, Judgment of March 29. www.corteidh.or.cr/docs/casos/articulos/seriec_146_ing.pdf.

Ioannidis, Konstantinos. 2019. "Heterotopic Landscapes: From GreenParks to Hybrid Territories." In *CyberParks – The Interface Between People, Places and Technology: New Approaches and Perspectives*, edited by Carlos Smaniotto Costa, Ina Šuklje Erjavec, Therese Kenna, and Michiel de Lange, 14–24. Cham: Springer.

Jameson, Fredric. 1994. *The Seeds of Time*. Columbia: Columbia University Press.

Jessop, Bob, Neil Brenner, and Martin Jones. 2008. "Theorizing Sociospatial Relations." *Environment and Planning D: Society and Space* 26 (3): 389–401.

Johnson, Peter. 2013. "The Geographies of Heterotopia." *Geography Compass* 7 (11): 790–803.

Klein, Naomi. 2015. *This Changes Everything: Capitalism vs. the Climate*. New York: Simon and Schuster.

Kleinpenning, Johan Martin Gerard. 1987. *Man and Land in Paraguay*. Providence: FORIS Publications.

Knight, Kelvin T. 2017. "Placeless Places: Resolving the Paradox of Foucault's Heterotopia." *Textual Practice* 31 (1): 141–58.

Lefebvre, Henri. 1991. *The Production of Space*. Translated by Donald Nicholson-Smith. Oxford: Blackwell.

Lund, Susan, James Manyika, Jonathan Woetzel, Jacques Bughin, Mekala Krishnan, Jeongmin Seong, and Mac Muir. 2019. "Globalization in Transition: The Future of Trade and Value Chains." *McKinsey Global Institute*. Accessed March 21, 2019. www. mckinsey. com/featured-insights/innovationand-growth/globalization-in-transition-the-future-of-trade-and-value-chains.

Martin, David A. 1999. "Building Heterotopia: Realism, Sovereignty, and Development in the Ecuadoran Amazon." *Alternatives* 24 (1): 59–81.

Moore, Donald S. 2005. *Suffering for Territory: Race, Place, and Power in Zimbabwe*. Durham: Duke University Press.

Moore, Sally Falk. 2000. *Law as Process: An Anthropological Approach*. Münster: LIT Verlag.

Neupert, Ricardo F. 1991. "La Colonización Brasileña en la Frontera Agrícola del Paraguay." *Notas de población* 51–52: 121–54.

Nickson, Andrew. 1981. "Brazilian Colonization of the Eastern Border Region of Paraguay." *Journal of Latin American Studies* 13 (1): 111–31.

Otsuki, Kei. 2012. "Illegality in Settlement Heterotopias: A Study of Frontier Governance in the Brazilian Amazon." *Environment and Planning D: Society and Space* 30 (5): 896–912.

Palladino, Mariangela, and John Miller, eds. 2015. *The Globalization of Space: Foucault and Heterotopia*. Abingdon: Routledge.

Pendleton, Michael R. 2007. "The Social Basis of Illegal Logging and Forestry Law Enforcement in North America." In *Illegal Logging: Law Enforcement, Livelihoods and the Timber Trade*, edited by Luca Tacconi, 17–42. London: Taylor & Francis.

Pietrobelli, Carlo, and Roberta Rabellotti. 2007. "Business Development Service Centres in Italy: Close to Firms, Far from Innovation." *World Review of Science, Technology and Sustainable Development* 4 (1): 38–55.

Quinn, Bernadette, and Linda Wilks. 2017. "Festival Heterotopias: Spatial and Temporal Transformations in Two Small-scale Settlements." *Journal of Rural Studies* 53: 35–44.

Saldanha, Arun. 2008. "Heterotopia and Structuralism." *Environment and Planning A* 40 (9): 2080–96.

St John, Graham. 2001. "Alternative Cultural Heterotopia and the Liminoid Body: Beyond Turner at ConFest." *The Australian Journal of Anthropology* 12 (1): 47–66.

Taussig, Michael. 2000. "The Beach (a Fantasy)." *Critical Inquiry* 26 (2): 249–78.

Taylor, Dianna, and Joanna Crosby. 2018. "Introductory Essay: Foucauldian Spaces." *Foucault Studies* 24: 6–11.

Turner, Victor Witter. 1982. *From Ritual to Theatre: The Human Seriousness of Play*. New York: PAJ Publications.

Turzi, Mariano. 2017. *The Political Economy of Agricultural Booms: Managing Soybean Production in Argentina, Brazil, and Paraguay*. London: Springer.

Venkatesan, Soumhya. 2009. "Rethinking Agency: Persons and Things in the Heterotopia of 'Traditional Indian Craft'." *Journal of the Royal Anthropological Institute* 15 (1): 78–95.

Watts, Michael, and Richard Peet. 2004. "Liberating Political Ecology." *Liberation Ecologies: Environment, Development, Social Movements* 2: 3–43.

6 "Interesting and incompatible relationships"

Force and form in Pedro Costa's porous city

Adam Kildare Cottrel

Pedro Costa's *Ossos* (1997) is the first of three films documenting the space of Fontainhas, a network of improvised housing on the outskirts of Lisbon, its eventual demolition, and the expulsion and relocation of its residents to a public housing project. This chapter's focus is on the first entry, which concerns the intermingled lives of three wayward youths struggling to survive in the turbulent urban landscape. The narrative revolves around a young woman named Clotilde (Vanda Duarte), her best friend Tina (Mariya Limpkin), and Tina's lover (Nuno Valdez), who are recent parents to a child they cannot provide for. The situation is dire from the outset. Tina and the child's father look to be teenagers, almost children themselves, spending their days idling in their makeshift house and travelling by bus and foot to more affluent parts of the city, like the homes of Lisbon's middle class where they work as day cleaners. Not twenty minutes into the film, Tina attempts to kill herself and her child by releasing a drum of natural gas into their cramped living quarters. The child's father, upon returning home and piecing together what is happening, takes the child, stuffs it in a plastic shopping bag and exits out onto the bustling city streets. From here things become even more troubling with the father traveling by foot through various pockets of Lisbon looking to sell the child to someone who can provide cash on delivery.

If we were to judge Costa's film based purely on its story, we might very well believe it to be a melodrama, replete with entangled love lives, unwanted children, and numerous crimes that straddle the line between passion and desperation. And yet, for all of its ostensible drama, the tone is reserved and the story seems secondary to the film's true star: the space of Fontainhas. This point of observation becomes apparent quite early in the film, as Costa's camera lingers on rooms draped in streetlight, or corners shrouded by shadow, but otherwise without activity or narrative significance. Establishing shots of exterior locations are prominent yet ambiguous, the film repeatedly introducing specific places from one point in space only later to reintroduce them from another, rendering the structure nearly unrecognisable. Presented as such, the space of Fontainhas has the tendency of folding back in on itself, trading geographic certainty for qualitative contemplation.

What are we to contemplate? In this too the film is initially ambiguous due to its minimal, wandering presentation. While *what* is happening may seem at home in a dime-store novel, *how* we are seeing it keeps our focus elsewhere. Most prominently, the film's drifting narrative structure elevates the sensory experience, lingering on sights, sounds, and actions to highlight their primary importance. For example, in countless scenes we hear sounds organic to the makeshift homes, such as a leaking faucet or creaking wood floors, rub against the street noise from the city traffic, like honking horns or roaring engines. There is also ample screen time given to laboured and fatigued bodies, granting observational purchase to the differences and similarities between white- and blue-collar work and lifestyles. These moments are striking, for instance, when we see our young protagonists showering in the townhouses they clean, in order to take advantage of the opportunity to use indoor plumbing. Or later, when a local nurse tries to help by visiting Fontainhas in her radiant white linen uniform, creating a visual schism that resonates with the socio-economic disparity.

We find additional juxtapositions regarding speed and travel made prominent with Costa's serial use of long takes, creating "action" sequences that can take minutes to unfold. These sequences delay and frustrate, never quite reaching a crescendo, instead, prolonging the duration of movement and the effort to exercise it. This plays out most noticeably when the father travels by foot carrying his newborn child in a plastic bag through the streets of Lisbon. The long, uninterrupted shots dramatise and magnify the physical act of walking, all the while juxtaposing his laboured movement with the deft ease of passing cars speeding by him. Anyone who has seen this film would hesitate to call sequences such as this a montage – its pacing too glacial and its camera work too adrift to exact certainty from the visual composition. In these lingering moments, though, Costa expands the duration of each shot so we can better absorb the atmospheric forces that pervade. The slow, dawdling moments serialised in the film offer us opportunities to witness a landscape that feels adrift from the surrounding commercial standards of urban living. These spaces are not initially visible but become increasingly so over the course of the film, and include a public bus, a hospital, urban apartments, and a city square.

These specific places oddly stand out in *Ossos* for their normalcy, complacency, and conformity to late twentieth-century standards of Western living visibly and sometimes violently clashing with the organically unordered nature of Fontainhas. In fact, Costa's distended pacing makes the familiar places of city, state, and commerce almost alien, as the film repeatedly asks us to engage and dwell for longer periods of time in the alternative paces, expectations, and social formations of Fontainhas. As such, Fontainhas resembles what Michel Foucault calls heterotopia, or space that disturbs, intensifies, contradicts, or transforms the predominant landscape. What is of interest to me here is how *Ossos* captures social relations in the midst of spatial change, in this case the gentrification of the Fontainhas area of Lisbon. By emphasising the social relations of its young protagonists, the film importantly gives insight into what heterotopia might mean in the midst of a global change in living standards.

It is important then as we explore the representational and presentational nature of *Ossos* that we keep in mind Foucault's defining proposition for heterotopia as one node amongst many in a specific historical constellation. That is, for as much as his essay is about exploring alternative notions of spatial and temporal logics, he is equally sensitive to how any one heterotopia is historically constructed. After all, Foucault proposes first that his own ruminations on individual places have led him to ponder not just space but "the epoch of space" (Foucault 1986, 22). In this, we might do well to consider our own epoch as we continue to engage with heterotopia, which we know to be one increasingly defined by global perspectives, scales, and connections. Fontainhas provides an interesting test case here, a place that is both unique due to its organic ordering but also vulnerable to be reordered because of it. While what necessarily defines heterotopia is contingent upon any given time or any one place, the film's repeated juxtapositions between Fontainhas' and Lisbon's commercial districts help give semblance as to what defines alternative space in the shadow of globalisation.

Here, I am proposing that we take seriously how *Ossos* documents Fontainhas, not simply as an example of heterotopia, but how the formal presentation's repeated tendency to congeal a multiplicity of people and places, cultures and economies already gives us an idea as to what a globally integrated landscape might entail. Within this visual constellation the film gives us access to a moment in history that captures part of Lisbon in transition from the archaic shantytowns of Fontainhas to modernity's promise of better living through gentrification. In doing so, we are provided with an optic by which we can see how the transformation of space determines its capacity to permit certain social relations and deny others. In this process we are able to glimpse some telling signs of heterotopia in a way very similar to how Foucault describes it, as "an ensemble of relations that makes them appear as juxtaposed, set off against one another, implicated by each other – that makes them appear, in short, as a sort of configuration" (Foucault 1986, 22).

Ossos presents us with a variety of configurations, highlighting how divergent elements are separated or entwined based on spatial logic and proximity, creating the potential for social relations in some cases and denying them in others. The film's repeated emphasis on spatial shifts and visual disparity is revealing for how it cultivates an awareness that the discrepancy between spaces stems from influences that are present but not in any obvious way. These influences or abstract powers are what I will identify here as *force*, which can be found in many forms but also does not necessarily carry form. When we think of force we most readily think of bodily actions and reactions, such as physical labour or even a violent encounter. But *Ossos* helps clue us in to an additional conception of force that influences, directs, and even determines lived life yet remains allusive to our tangible experience of it. Here we can think of how certain social and cultural expectations dictate our day-to-day actions or performance. Likewise, we can think of how labour standards or political policies exert influence and discipline without ever raising a fist or delivering a blow. Force, in this conception, is defined by its abstraction, which often makes it hard to pinpoint because it only resonates secondarily. While force may seem like an abstract concept, it is the hard sciences where we can find one of its most helpful definitions. Kriti Sharma writes:

"'Force' in some contexts seems to mean 'what keeps things together' (or perhaps 'what keeps things apart'), but more precisely means 'what helps predict whether things will be observed together'" (Sharma 2015, 30). Here, force is a power that can bind, separate, or change entities in relation to one another.

For Sharma, force is an influencing power that contingently entwines or separates entities, which helps us to see how the absence or presence of one entity can come to define another. These entities can range from molecules to humans, but they are defined by their capacity to aggregate or segregate based on spatial proximity. This notion of force is helpful as we unpack the landscape of Fontainhas, as Costa's film dramatises how disparate elements find each other, or don't, and what that might tell us about the defining markers of this alternative space. As we will see in this film, the growing economic disparity between Lisbon's middle- and working-class creates stark and contrasting juxtapositions – socially, economically, and aesthetically. In this, I see evidence of how heterotopias in today's global landscape seem to be increasingly defined by juxtapositions, and how this film helps us to understand how the privileged and the abject are positioned in relation to one another.

In consistently bringing entities together, we can observe juxtapositions that help give clues as to what forces may be in play, and what this means to live in spaces defined by them. In this chapter, I argue that the recurrent emphasis on temporal and rhythmic difference, highlighted by the film's decelerated style, offers us images that reveal the prevalence of force in spaces defined by juxtaposition. While force may not always be directly observable, the juxtapositions Costa's films produce help render them visible, giving us impressions that influence the physical and material bodies we witness. We can think of force as uniting entities, such as the collection of immigrants and refugees who populate Fontainhas, just as we can equally imagine it separating those living in the shantytowns from the middle class.

The film's decelerated style thus offers a counter rhythm from the normative pace of life, helping to sustain a temporal experience where forces that infiltrate, influence, and, ultimately, remake spatial logics are rendered tangible in the visual disparity of the images. With this aesthetic choice, *Ossos* helps us realise that Fontainhas is a space peculiar and its own, a space that requires patience and observation of its unique rhythm and sensibility. What we are to make of this formal tendency is that the film's slowness is part of its gambit. That is, the drifting quality it assumes expands our capacity to think spatially by dilating the cinematic experience of time. For in slowing down the narrative progression of the film to the point where it struggles to maintain principal importance, it opens up an aesthetic encounter where we must first learn to observe and dwell in the specificity of space.

Costa has stated that what initially drew him to Fontainhas – its specifying quality, as it were – was its capacity to tolerate a multiplicity of sights, sounds, smells, and social interactions. He explains:

> When I entered the Fontainhas area, there were colors and smells that made me remember the things and events of the past, and also ideas about people to which I am attracted. These ideas nestled close to each other, living together

even as they led very solitary lives because of violent and painful separation. A *form* of interesting and incompatible relationships existed in this.

(Costa 2001, my emphasis)

Based on Costa's description the relationship between Fontainhas and heterotopia seems most appropriate in reference to Foucault's third principle, which states that "heterotopia is capable of juxtaposing in a single real place several spaces, several sites that are in themselves incompatible" (Foucault 1986, 24). In *Ossos* we see how the attention given to juxtaposition sets the stage for my own concerns related to how forces of power make over, change, and discipline most egregiously those who are living in alternative spaces and temporal modes.

Because of the traumatic turbulence such structural disparity can have for the vulnerable, the impoverished, and the working poor, urban centres like Lisbon play a privileged role in helping to exact the consequences of an increasingly segregated commons. Costa's patient camera allows us to witness these issues through spatial transformation, unsettling and destabilising the ground where human subjects and bodies are constituted. With *Ossos* we get a sense of the ubiquity of crisis today, driven by rhythmic disturbances that segregate people into distinct populations concerning material wealth and personal mobility. These distinct populations too need their distinct spaces and much of what the film captures is how space defines our capacity to function in everyday life and even how space defines what capacities are available to us.

Herein lies the issue of why force and form matter to this volume's discussion on heterotopia. In offering a decelerated aesthetic, Costa is able to magnify the interdependent nature of Foucault's initial observations on alternative spaces. His observations, for instance, that these other spaces also necessitate their own temporal mode and that heterotopia is largely defined by a space's capacity to "arrive at a sort of absolute break with their traditional time" (Foucault 1986, 24), are certainly worth further reflection. Because what traditional time meant for Foucault and the heterotopias he articulated do not apply in all cases, especially when temporal structures are increasingly defined by forces that are the by-product of global power structures. This chapter, then, tries to put these points into dialogue analysing the disparate nature of lived life stemming from forces that disrupt *stabilising life patterns*, and the precarious effect these disruptions have on *spaces of living*. By attending to the rhythms of Fontainhas and the forces that penetrate and ultimately remake its space, we can better discern how logics of power segregate people into distinct populations defined by their capacity to absorb these impacts from those who cannot. Style here renders this space such that it "takes for us the *form* of relations among sites" (Foucault 1986, 24, my emphasis). Let us now turn our attention to these sites and what forms they offer to us.

Temporal desynchronisation

The opening minutes of *Ossos* introduce the space of Fontainhas by asking viewers to disengage from their own. The film starts with a long take of a young,

Figure 6.1 Ossos (1997), director Pedro Costa

unidentified woman (Zita Duarte) starring soberly into the camera (Figure 6.1). We do not know who this character is nor will we learn anything more of her life as the film progresses. As the young woman stares blankly back at us, our eyes and mind begin to wonder. We scan the screen for clues that might help us discern her identity, or the camera's interest in her. The shots are tight, framing her from the waist up sitting on what we can assume is a bed. The introduction to the space of Fontainhas from its interior, as opposed to an establishing shot of the exterior, attests to this place as both lived and personal. This choice asks us to understand the surface space of Fontainhas as a specific place, which gives important context to how the film is portraying the makeshift exterior. Instead of merely presenting the pirate-urbanisation of the shantytown, we must first consider the lives of those who occupy it.

The distended opening feels exorbitant when we try to negotiate the time endured watching what seems like not much at all. But in these shots Costa's camera makes a statement about knowledge regarding heterotopic space: principally that such knowledge is to be gained through the lived experience of it. Because of this there is a palpable sense that the screen we are viewing acts as a barrier, a boundary that can only be overcome through a gradual immersion of the space depicted. For example, the film's cautious opening denies us a master shot that would give us immediate spatial orientation. Instead, space is presented so as to deny us mastery, prompting us to patiently explore the entirety of the screen so we might come to learn its specificities and its particular orientation. In this instance, the long take asks us to consider what we know or can know about a place and people without first occupying the given terrain. We can derive here that space is not meant to be immediately knowable to outsiders and that knowledge must be accumulated over time.

Costa's camera functions as a pedagogical device by devising a temporal regime that provides ample time to sit and contemplate the entirety of the screen. With this time spectators are able to more fully immerse themselves into the space, orienting to the rhythm and pulsations of its inhabitants. The camera does not attempt to forcibly connect the viewer to its on-screen counterpart with this dilation of time. Instead, it places us within proximity, allowing us to co-exist in space. For Lutz Koepnik, the power of the long take is found in exactly these terms, that is, for its capacity to establish a temporality that asks us to think rather than simply presenting us with a causal chain of narrative information. He writes:

> [W]e cannot overestimate the political import of the long take's evocation of patience and waiting, of sustained seeing and listening, of the nonintentional and wonderous, of its strategies of framing and unframing time so as to inhabit space as something whose complexity we can never fully map and comprehend, let alone own and master.
>
> (Koepnik 2017, 221)

For Koepnik, our access to engage and think with alternative space is provocatively aided by film's capacity to stage moments that complicate our preconceived notions of it. Space, in this instance, becomes strange and complex, eluding mastery through complexity the longer we dwell within its coordinates.

At the same time, if we return to these initial images and the space made available to us by this uncoiling of time, we cannot escape the young girl and her forlorn face. The felt proximity brings to the surface not only what we are seeing but also the question of how our space might look to her. In this sense, the long take functions like a mirror, reflecting one image of lived life so as to prompt reflection on our own. In this reflection the long take asks us to detach from the temporal logic of our lived life so as to experience it cinematically. The time to contemplate our surrounding allows us to ponder space alternatively from how we expect to experience it. This mirroring function is reflexive in that the viewer must come to terms with the learned knowledge of their space and the temporal logics that define it. For example, we might initially feel frustrated or even confused at the measured pacing of the film and the perceived lack of action in Fontainhas as it so drastically contrasts with the normative parameters of capitalist culture. But over time these observations prompt our thoughts to our own space – one that is dominated by accelerated and intentioned movement, as opposed to the drift and delay we find in Fontainhas. The juxtaposition between the two inaugurates our own initiation into Fontainhas. The mirror, therefore, is a starting point of departure that first maintains that we reflect and recognise the specificity of our space, provoking constant reflection between self and other.

Mirrors also direct the reader of Foucault's essay on heterotopia regarding how we might engage with spaces that are other from our own, or other from what is normative for a given social order. "From the standpoint of the mirror," Foucault writes,

> I discover my absence from the place where I am since I see myself over there. Starting from this gaze that is, as it were, directed toward me, from the ground of this virtual space that is on the other side of the glass, I come back toward myself and to reconstitute myself there where I am.
>
> (Foucault 1986, 20)

For Foucault, the mirror stages a moment of self-reflection where the observer is asked to consider not simply the space of the local subject but spaces that occupy a single place, both where we stand and where we can imagine ourselves standing as reflection. Costa's work does something quite similar in reproducing these mirror-shots, not just in the film's opening minutes but repeatedly. These reflexive meditations tie the spatial to the personal, prompting viewers to make connections amongst disparate actions and places, even as the film's narrative denies such concrete connections. Heterotopias are, in this sense of the word, worlds within worlds, mirroring and yet upsetting what is outside, while at the same time juxtaposing one's present location with those occupied by unknowable others.

Ossos requires our time and patience in order to cultivate a sensitivity to what makes Fontainhas not simply different, but also defiantly its own. If the film's only interest in space were Fontainhas, then it would be more than appropriate to catalogue its traits, reflect on and assign its attributes to certain categories and conditions. But its inhabitants are not bound to this one locale and its organic pace of life. Part of our education concerning heterotopia is made possible by the multiplicity of spaces our protagonists must negotiate on a daily basis. We first exit Fontainhas soon after leaving the unnamed girl in her room. The film's next shot sequence reconstructs the semblance of a working day for Clotilde, Tina, and her lover. Upon exiting the borders of Fontainhas, our young protagonists find themselves engulfed in a variety of temporal modes dependent on the spatial logic of the given location. *Ossos* puts into formal play one of its most revealing points concerning heterotopia in our age of globalism here, painting a globalised landscape as one that encourages conditions where time no longer follows one singular pace or even belongs to any specific space. As subjects of global forces, the film suggests, we are compelled to live lives that treat time and spatial orders as multiple and competing, simultaneously local and global, progressive and circular.

Our first image outside of Fontainhas is of Tina and her lover waiting at a public bus stop. The outside world is hard to discern, Costa's camera magnifying its subjects in close-up, which helps reinforce their own sense of caution and uncertainty about the outside world. The scene holds us in anticipation and when it's apparent we are waiting for nothing more than the bus, it begins to drag, asking we sacrifice our time to wait with our on-screen counterparts. The power of such a simple juxtaposition establishes a spatial transition that bifurcates two distinct temporal modes, one related to the pacing of the film and one regarding our own patience in viewing. This process demands that we as viewers undergo a similar transition. A moment like this helps place the relation between space and speed at the crux of the individual's constitution of self by mediating its orientation to the environment.

For example, the proceeding shot continues to hold these paces together in a productive tension, framing the characters' interlaced hands in a close-up as they sit idle on the city bus. Simultaneously the roar of the bus's diesel engine saturates the scene. The bus, an automated form of accelerated movement, clearly signals we are collectively being brought "up to speed," into Lisbon's upper middle-class neighbourhoods. There is, if only fleetingly, a sense of anticipation this mechanised speed induces. It feels as if we might be going "somewhere" if not exiting "nowhere" and its accompanying logic and rhythm.

The camera then abruptly cuts to a serene white door, prominently displaying an eyehole and four security locks. As the teens enter the townhouse they are about to clean, we cannot escape the contrast in settings. The images radiate with luminescent white walls, furnishings impeccably kempt, and spaces clearly demarcated for a living room, bedroom, kitchen, and closet. The space of this middle-class home when compared to Fontainhas seems hermetically sealed from outside disturbances, sounds unwanted, intrusions unwelcome. There is a real sense that the room is designed to prohibit exposure from exterior forces, its deafening silence providing a quiet space for recuperation and leisure. By comparison, Fontainhas is open and vulnerable to the outside world and all the intrusions that come with it: traffic noise, private conversations of neighbours, and unwanted light are all a by-product of its open corridors, entranceways, and windows. Despite its open design the camera remains fixed in close-up, giving little if any sense that there is much room to manoeuvre. In contrast, the townhouse is shot so as to highlight the expansiveness of the space. This is highlighted by the camera's placement, which works to present the space of each room through the frame of another. For instance, the camera is positioned in a room adjacent to the living room so that it is framed by the walkway to the dining room, or the bathroom by the open doorway to the hallway.

The sense this space gives is that it is both conducive to the freedom and flow of movement with its open style while still guarding against the ceaseless activity of the shared, public spaces of the city. Space works here as a barrier to outside forces and influences, or at least a reprieve for those who can afford it. This is space contoured for the individual, the entrepreneur, the self-mobile and self-sufficient subject of the global marketplace. Security and mobility are tethered in this spatial construction, prompting us to reflect back upon the rather porous and communal nature of Fontainhas. Further, this spatial juxtaposition highlights the ongoing desynchronisation that such stark contrasts in living suggest, where structural couplings between places disrupt social solidarity and our capacity to engage and relate to others.

The multiplicity of temporal modes determined by the various spatial logics may at first seem innocuous, but as the second and third instalments of this trilogy will attest, falling out of pace with the dominant temporal order can have lasting and dire consequences for those who resist or falter under these demands. Being out of pace, as this sequence shows us, often means being outpaced. Suggested here is that surviving this landscape is strongly tied to negotiating its multiplicity of temporal rhythms highly dependent on a given space's organisational logic. The disadvantaged social status as precarious youths places Tina and her lover in a situation where they must learn to navigate not only their own environment but also those of others. The film suggests something quite dire here about heterotopia, raising the question of what is "other" or "alternate" in an era where living practices are globally homogenised to a middle-class standard of consumer culture. For the young protagonists, cleaning in the open and luxurious space of the townhouse makes manifest the simultaneity of the non-simultaneous, elevating the notion that desynchronisation is a by-product of spatial logics that either conform to normative standards or fall out of step with them.

These images make clear that maintaining pace becomes its own form of exhausting labour. The protagonists, despite their best efforts, struggle under this standard, implicitly realising that standing still in a globalised world almost always means falling behind. For political theorist Franco Berardi, the freedom promised by the contingency of the marketplace has led to inhuman constructs of space and time, which have in turn leveraged temporal forces, like increased pace of living, as a weapon that can inflict violence on those unable to absorb its impact, people and space alike. "The global deterritorialization of financial capitalism," Berardi writes, "has spread precariousness, psychic fragility, and desolidarization. Therefore the current precarious insurrection questions the rhythmic disturbance provoked by semio-capital, and tries to overcome our existing inability to tune into a shared vibration" (Berardi 2012, 37). His concern with our inability to establish a "shared vibration" relates to how force separates the commons into distinct market shares, segregated by their individual pulse in the larger rhythm of social and economic life. We can understand force in this particular instance as a mediated rhythm of life, particularly how capital's penchant for flow and flexibility alters the way we live and how we understand the experience of it. For Berardi, the lived experience of time, in this case the rhythm of everyday life, separates certain populations into distinct spatial locales, elevating difference and disparity at the expense of social solidarity. Our inability to help each other, Berardi would argue, is largely a by-product of our inability to experience life collectively, meaning, at a singular or normative pace.

Berardi's observation is one reason why Costa's deliberate slowness is of value; it formalises a "shared vibration" by mediating the temporal experience of the film. In doing so, the rhythm establishes a collective temporal logic, which helps us experience how space and social solidarity are fractured and separated from each other. We can think back here to Tina and her lover waiting for the city bus as a prime example. This shared experience of waiting certainly creates a spatially proximate connection. But for those living in the townhouse they clean, it would be hard to imagine a similar encounter. Queuing in line, as we can attest to in so many situations today, is a province for the underserved and underprivileged. It is a spatially inflected inconvenience that can all too easily be avoided with bourgeois consumer distinctions such as first-class ticketing, or an EZ-pass platform for turnpike travel. Randy Martin argues that this social segmentation is a principal component of our spatial orientation, going as far to suggest that "the binary to have emerged over the past thirty years would separate the risk-capable from those considered at risk" (Martin 2007, 136). In this particular rendering, social rhythms have a growing influence over our capacity to unite and solidify, as well as divide and segregate, making the line between those risk-capable from those at risk all the more problematic.

Spatial segregation

Costa's patient, foreboding camera work helps define the capacity for decelerated aesthetics to document the seemingly minute in order to present its often obscured and overlooked significance. We saw this in the longing stares from the unnamed girl and how those images appeal to a contingent, reflexive spectatorship, as well as the purposeful yet unhurried shot sequences that framed the desynchronous nature between Fontainhas and the townhouse. Visual juxtapositions like these are the by-product of the camera's capacity to hold in productive tension a myriad of elements simultaneously. These images give a sense about the appeal this film is making regarding Fontainhas and heterotopia. Specifically, how forces holding disparate elements together both define heterotopia in this particular case, but also how globalisation itself is increasingly defined under these same terms. Put another way, the film shows us how global landscapes aggregate huge swathes of the human population as urban centres increasingly define the economic viability of space. Social orders organised in this way must entertain people and customs from a large array of humanity, aggregating difference and thus opening the possibility to a multitude of encounters. On one hand, we find this to be a normative logic of how space is used in a global city; on the other hand this practice increases the contingent encounters and possibilities, facilitating spaces that drift away or deviate otherwise from the norm.

This might seem like a contradiction; after all one of the founding tenants of globalisation is the anxiety about the homogenisation of space and culture, where difference and specificity succumb to the forces of a universal standard. We can think here how prevalent Western, capitalist constructions of space dictate land as private property. Or how our notion of the good life is almost always tied to the capacity to participate in a bourgeois consumer culture. But this apparent antagonism is not far off from how Fredric Jameson introduced his own influential conception of globalisation, defining it as an "untotalizable totality which intensifies binary relations between its parts" (Jameson 1998, xii). Most pointedly, we find how the intensification of relations is predicated upon force's capacity to aggregate or segregate entities in space. As we know, first-world conceptions of life are materially impossible without an exploitable third-world population. But what we see in *Ossos* is how the first- and third-world, while separated in lifestyle, are increasingly aggregated in space. The film gives representation to these entities that are spatially proximate but increasingly disparate, giving us images that dramatically stage those living side-by-side yet in entirely separate social realities.

In *Ossos* we do not yet see the literal, material destruction of Fontainhas, but we do see how the intensification of binary relations between parts can create disparities between newly interdependent entities. The result of force's capacity to bind through spatial proximity is that those already at risk become increasingly disposable – their organic temporal and spatial conditions register as negative drag to the forward propulsion of capital. These disjunctions can be understood as speed bumps in an increasingly commercialised landscape that privileges speed, flow, and liquidity. Regarding the constitutional role of force to lived life, Giovanni Arrighi stated that at the end of the twentieth century, with the emergence of financial capital as the global ethos of power, society must adhere to "the centrality of 'force' in determining the distribution of costs and benefits among participants in the market economy" (Arrighi 1994, 19). Part of this procedure, Arrighi argues, must attend to how space is being reconstituted from a "space-of-places" to a "space-of-flows" (Arrighi 1994, 23). For Arrighi, space under the dictates of capital is an attempt to accommodate easier and faster channels of circulation. The transition from a space-of-places to a space-of-flows directs our attention not only to the lifestyles that will thrive in and with flow, but also to the kinds of ideas, relations, and geography that must be transformed, updated, or expelled to maintain it.

Ossos is paramount to our understanding of this point, as we are given access to Fontainhas almost entirely in compartmentalised shots that suggest a privileging of place over flow. We see pockets of the pirate urbanisation, we witness small moments, and we listen to conversations happening off-screen. As such, the film presents Fontainhas as incomprehensible as a totality, with no identifiable social order or universal communal standards. This is reinforced in the tight, close-up shots, returning at several times to a high angle, canted image of the shanty's rooftops to illustrate the lack of geometrical design. Space in Fontainhas is congested, improvised, and constantly changing due to its permeable state. This permeability hints at how vulnerable Fontainhas is to forces that reorient space and time, with its open winding walkways, its disregard for demarcation, its lack of boundaries and discernible border.

Ossos is dotted with seemingly innocuous moments that illustrate the open, labyrinthine quality of Fontainhas. For instance, the film's halfway point stages a dance party in a makeshift hall. We find Tina and Clotilde standing on the perimeter of the dance floor, initially hesitant to join the group. They huddle together as foot traffic crosses the screen from both the left and right. We hear the music echo out into the exterior corridors, we see the residents bob and weave to the beat, and feel a sense of communal gathering. As the scene progresses the camera repeatedly cuts between Tina and Clotilde as smiles break on their faces, the muscles around their eyes relax until eventually they become integrated with the group. The camera then cuts to a medium-shot, revealing a mix of Portuguese nationals and East African immigrants from Portugal's former colonies. The room is crowded and unadorned, its ceiling swayed, we watch transfixed by this collective rhythmic pulse.

Costa carefully frames these shots in order to reveal one or more rooms from its point of view, giving us a sense that this space is more like an interconnected hive than standalone structures grouped together. As the dance reaches its fervid pitch the image abruptly cuts to Tina passing out either from exhaustion or chemically induced stupor, hitting the floor with an audible thud. Then, another juxtaposition, as the camera cuts again to an older African couple dancing down a narrow corridor before finally cutting again to a series of cars passing by the exterior side of the building that faces the street. With this sequence, Costa brings to our attention bodies in space that create communal belonging. These bodies move and gyrate to a rhythm defined by this place, but the scene is forcefully punctuated by an alternative pace, one defined by the mechanical speed of the automobile and its decidedly inhuman rate of movement. In this juxtaposition, the film gives form to the porous quality of Fontainhas, which speaks as much to its capacity to stage a communal block party as it does to its antiquated and vulnerable design.

In his essay "Naples," Walter Benjamin poignantly discusses urban space evoking the central image of porosity. I want to suggest that what Benjamin wrote about Naples is equally illustrative of Fontainhas, especially in regards to the relationship between spatial design and lifestyle. He writes:

> As porous as this stone is the architecture. Building and action interpenetrate in the courtyards, arcades, and stairways. In everything, they preserve the scope to become a theatre of new, unforeseen constellation. The stamp of definitive is avoided. No situation appears intended forever, no figure asserts it "thus and not otherwise." This is how architecture, the most binding part of the communal rhythm, comes into being here.
>
> (Benjamin 1986, 165–66)

Benjamin's porosity emphasises how space is open and conducive to social gathering and communal action, allowing space to take on a variety of forms and functions contingent to any given moment. But while its openness to contingency may help define its specificity, and its capacity to bind entities, it is also vulnerable to forces designed to segregate and streamline. This porous space is simultaneously a ground of bliss and despair. Its openness to fortuitous encounters and the "non-productive" time of leisure also opens it to forces prone to transform it from a specific place to a conduit of flow.

Space's capacity to induce flow is best exemplified in one of the few exterior scenes in the film. Here, the newborn's father has taken his child to a busy city square in hopes of securing food or money (Figure 6.2). The camera is static, positioned at eye level framing the father walking in small semicircles, doubling back every few meters. His movements are restricted by the surge of people hustling in and out of the frame moving with the flow of traffic. In contrast, the father's movements do not align with the flow of traffic, his body swaying side-to-side hoping to interrupt this mass of human movement by siphoning off one from the group. The space in question is framed to showcase its architectural elegance, sophisticated network of bookshops, old-style cafes, and art nouveau jewellery shops, highlighting the straight lines, conducive to moving through and into its commercial venues.

Figure 6.2 Ossos (1997), director Pedro Costa

At one point the father attempts to engage a large man in a navy blue business suit as he enters the shot and stops dead centre in the frame. He stands a few feet closer to the camera than most of the pedestrians walking by, dominating the frame, and eclipsing a mammoth vertical pillar that sets in the background. The shot is framed so as to create a vertical line that runs from the top of the pillar to the centre of the square. The buildings on either side serve to further embellish this vector, creating a straight path where we find the force of speed flowing through this commercial district. The father asks the businessman for help, without making eye contact he shakes his head in a dismissive manner, looks away, and proceeds to re-join the flow of human traffic down the street and out of shot. In his absence the entrance to the Lisbon Metro, a citywide subway system, appears before us, cementing flow as the governing logic of the mainstream.

Here we see how a space-of-flows, with its emphasis on streamlined movement, prohibits the types of communal encounters found in Fontainhas. Further, the optic alignment between the pillar, the businessman, and the subway system reinforces the metaphorical alignment between these symbolic agents of governance, economy, and infrastructure. Clearly, the space of the public square is presented to contrast with the network of interconnecting alleyways and side streets of Fontainhas. Questions of spatial construction, and the social life made possible by it, surface in formal juxtapositions such as these. Here, we get a real sense about what life looks like based on spatial design, allotting or denying, aggregating or segregating populations based on class, ethnicity, and capacity to keep pace with the demands of the global market place.

Costa's juxtapositions highlight the capacity for decelerated aesthetics to render visible the abstract and underlying logics that determine how value is spatially constructed. These juxtapositions give form to forces that contour life, effectively materialising abstract power so that it is counted as real. Films like *Ossos* help us understand life dynamics through form by helping us see relationally between one space or population and another. But Costa's aesthetic also reveals that we see in this way, making explicit what otherwise slides behind the flow of action. As Martin Brady has helpfully formulated this point in relation to decelerated aesthetics, "change demands attention to detail, attention to detail demands time, time demands slowness" (Brady 2016, 81). For Brady, and certainly for Costa, slowness in the form of decelerated aesthetics is what allows us to register these logics, so we can better judge what changes become desirable because what is present is so brutal.

Rethinking global heterotopias

Due to the pervasive nature of segregation and control that spatially induced social rhythms can impose, it is no surprise that we have seen with increasing regularity how social commentary and media entertainment are discussed as representing, responding, or exploring crises such as we have examined here. Regarding this point, Lauren Berlant has gone so far as to argue that these socio-aesthetic concerns have manifested a genre of crisis. She writes:

> [A]cross diverse geopolitical and biopolitical locations, the present moment increasingly imposes itself on consciousness as a moment in extended crisis, with one happening piling on another. The genre of crisis is itself a heightening interpretive genre, rhetorically turning an ongoing condition into an intensified situation in which extensive threats to survival are said to dominate the reproduction of life.
>
> (Berlant 2011, 7)

I will add here that images of crisis lend themselves to moments where space becomes increasingly scrutinised for its capacity to generate profit, often making it untenable to remain untethered to capital. Large parts of Costa's trilogy document the porosity of the older, slower space of Fontainhas as it struggles under the strain of modern expectations regarding travel, communication, and commerce. Ultimately, as we see in *In Vanda's Room* (2000), Fontainhas is demolished and with *Colossal Youth* (2006), its occupants are relocated to a frighteningly austere housing project. The images of violent destruction and isolating assimilation offer us indelible moments that help to shape this burgeoning genre of crisis. Each of these films, in their own way, tells us a story about what constitutes heterotopia in a global landscape.

The issue at hand for these films, ultimately, is not whether the experience of life is motivated or influenced by force, but rather how force defines spatial logics where life lives. Sarah Sharma isolates the common denominator found between the various forces that define the early twenty-first century in order to highlight this point: "No matter what the specific contours of capital are, whether we call it fast capital, neoliberalism, late capitalism, or empire, capital develops at the expense of bodies" (Sharma 2014, 17). Heterotopic space becomes a question in this moment when bodies (both corporeal and material) need to be recalibrated, updated, and modernised to fit into the new faster pulsations of the market that expand the terrain by which capital can extract profit. These films offer for our consideration the real challenge a global capitalist landscape poses to alternative spaces for how it violently dismisses what falls outside its slipstream.

Images of heterotopic space are important to consider, then, because they offer us the opportunity to sense and contemplate the kinds of entities that global forces aggregate and segregate. It can provide some clues, or even establish a heuristic, for how we can see space as a repository for logics of power that can determine not only where we live, but also what kind of life that space can accommodate. In a world increasingly defined by the vibration of capital's rhythm, alternative spatial and temporal forms can help rupture our expected encounters with the world, forcing us to negotiate new ideas, people, or places. *Ossos*'s decelerated quality helps us negotiate a spatial configuration like Fontainhas, where the expected pace of the everyday is slowed to a crawl, reconstituting our notion of what is and therefore what should be. This formal deceleration permits us not simply to see what a non-conforming place looks like, but to explore and experience its lived alterity. *Ossos* formally invites us into this space, asking us to see how we have been segregated while also illuminating our interdependence.

Works cited

Arrighi, Giovanni. 1994. *The Long Twentieth Century: Money, Power, and the Origins of Our Times*. London and New York: Verso.

Benjamin, Walter. 1986. "Naples." In *Reflections: Essays, Aphorisms, Autobiographical Writings*, translated by Edmund Jephcott and edited by Peter Demetz, 163–73. New York and London: Schocken Books.

Berardi, Franco. 2012. *The Uprising: On Poetry and Finance*. Los Angeles: Semiotext(e).

Berlant, Lauren. 2011. *Cruel Optimism*. Durham and London: Duke University Press.

Brady, Martin. 2016. "'The Attitude of Smoking and Observing': Slow Film and Politics in the Cinema of Jean-Marie Straub and Danièle Huillet." In *Slow Cinema*, edited by Tiago de Luca and Nuno Barradas Jorge, 71–84. Edinburgh: Edinburgh University Press.

Colossal Youth. 2006. DVD. Directed by Pedro Costa. New York, NY: The Criterion Collection.

Costa, Pedro. 2001. "Director's Statement." *Jornal de Letras*, March 7. www.yidff.jp/2001/cat021/01c031-e.html.

Foucault, Michel. 1986. "Of Other Spaces: Utopias and Heterotopias." Translated by Jay Miskowiec. *Diacritics* 16 (1): 22–27.

In Vanda's Room. 2000. DVD. Directed by Pedro Costa. New York, NY: The Criterion Collection.

Jameson, Fredric. 1998. "Preface." In *The Cultures of Globalization*, edited by Fredric Jameson and Maso Miyoshi, xi–xvi. Durham and London: Duke University Press.

Koepnik, Lutz. 2017. *The Long Take: Art Cinema and the Wonderous*. Minneapolis and London: University of Minnesota Press.

Martin, Randy. 2007. *An Empire of Indifference: American War and the Financial Logic of Risk Management*. Durham and London: Duke University Press.

Ossos. 1997. DVD. Directed by Pedro Costa. New York, NY: The Criterion Collection.

Sharma, Kriti. 2015. *Interdependence: Biology and Beyond*. Fordham: Fordham University Press.

Sharma, Sarah. 2014. *In the Meantime: Temporality and Cultural Politics*. Durham and London: Duke University Press.

7 Heterotopia and perspective

Towards a different imagining of landscape

Henrietta Simson

Introduction

As Malcolm Andrews argues, landscape painting in Western culture comes into its own during the sixteenth century, encouraged by the Italian fashion for designing villas so that views of the surrounding countryside become framed by the windows, thus "pictorializing [the] landscape" (1999, 56). In Palladio's famous Villa Barbaro, this "pictorializing" is enhanced by frescos of landscapes painted by Paolo Veronese (1528–1588) that complement and supplement the actual landscape views seen from the windows (Figure 7.1). Veronese's use of perspectival technique produces a vision that blends pictorial and physical landscape space, the actual and the painted melding into a sophisticated but apparently straightforward, rationally convincing image. Although these frescos were painted approximately 450 years ago, ideas about landscape have remained remarkably static since then, and it continues to be understood in visual terms – as a scenic background to human narrative – whether as a painted, filmed, or photographed image or as a physical landscape to be viewed or surveyed from a distance. In the next few paragraphs I want to analyse the complexities within this "pictorializing" before setting out an argument that understands the landscape image (and by implication the landscape itself) in terms that challenge its construction under the globalised visual assumptions of early twenty-first-century capitalism.

When Leon Battista Alberti first theorised the technique of linear perspective in 1435, his focus was to aid the painter in the "natural" depiction of objects and figures in order that they may make convincing stories ("*istoria*") rather than understanding perspective as creating paintings that depicted extensive space, which is how it appears 130 years later in Veronese's frescos and how it is more likely to be understood today (Alberti 2004, 67–78). Perspective now appears to describe the actual properties of space and this elision manifests clearly in the idea of landscape, and is reflected in the terminology: the horizon line, the dual vanishing and viewing points. Perspectival images appear somehow equivalent to both "natural human vision" and "objective external space," so that they seem to show their subjects in an unmediated, direct way (Mitchell 1986, 37). As such, they are susceptible to ideological power, and their relationship with Western paradigms is complex and far reaching. They have intertwined with capitalist structures since the Renaissance not least because, as Michael Baxandall has

Figure 7.1 Paulo Veronese, *Sala a Crociera*, 1560–1, fresco, Villa Barbaro, Maser

shown, the mathematical knowledge that propelled perspective's development was also an essential aspect of an increasingly sophisticated banking system. The formulas that were used by artists to measure and shape their paintings were the same formulas used by bankers and traders to calculate interest, convert currency, and to measure and cost their goods (Baxandall 1988, 96). Perspective is further linked to capitalist ideology by Martin Jay when he states that "the placement of objects in a relational visual field, objects with no intrinsic value of their own outside of these relations, may be said to have paralleled the fungibility of exchange value under capitalism" (Jay 1993, 59). And with the development of Alberti's "centric ray" into the concept of the horizon line, perspective came to facilitate the measuring, mapping, and therefore conquering of space – enabling European imperialism and capitalist global trade (Steyerl 2011).

Perspective is fundamental to Western visual structures, and its influence on the development of the concept of landscape cannot be overestimated. Space became understood in visual terms, while perspectival images appeared to describe the separation of Descartes's *res extensa* from *res cogito* through their dual vanishing and viewing points. In this way landscape became entrenched as "object," opposite to and utterly distinct from the "subject," which is defined as uniquely human. This is a double bind that fixed it as "nature" to our "culture" – passively understood as background to the foregrounded human – and also as "representation" – it is an image, formed in the rational mind and governed by the rules of perspective, and as such it problematically confirms the assumption of space that, as Doreen Massey states, exists as "stasis," rather than "heterogeneity" (Massey 2005, 19).

This visual predicament has been critiqued since the linguistic turn that emerged towards the end of the twentieth century, whereby language and the interpretation of signs became the cornerstone for thinking within the humanities and social sciences, and the study of the image as "sign" superseded previous Western notions of the rationally produced image of equivalence.[1] The deconstructive practices that were developed by prominent thinkers on the image from this time such as Norman Bryson revealed the ideological appropriation embedded within "realist" images, i.e. the assumption that they can somehow directly represent their objects without mediation. However, this interpretative approach has resulted in an erasure of sorts, whereby images (and for present purpose images of landscape specifically) are ultimately seen in linguistic terms. W. J. T. Mitchell's writing from 1986 reveals this methodology:

> The commonplace of modern studies of images, in fact, is that they must be understood as a kind of language; instead of providing a transparent window on the world, images are now regarded as the sort of sign that presents a deceptive appearance of naturalness and transparence concealing an opaque, distorting, arbitrary mechanism of representation, a process of ideological mystification.
>
> (Mitchell 1986, 6)

Understanding realist images as signs is clearly an important means of wresting them from ideological control; however, as Henri Lefebvre points out, the discourse that frames this approach (which included the writing of Julia Kristeva, Jacques Derrida, and Roland Barthes) continues to consign space to the mental realm and eliminates the social dimension from its characterisations (Lefebvre 1991, 5). Their thesis, Lefebvre claimed, "assumes the logical, epistemological and anthropological priority of language over space," a problem whereby "[t]he pre-existence of an objective, neutral and empty space is simply taken as read, and only the space of speech (and writing) is dealt with as something that must be created" (1991, 36). This linguistic approach continues to privilege the mental realm over the physical. It leaves Western spatial assumptions unchecked in the process, and therefore becomes an inherently problematic reformulation of the Cartesian *Cogito* (Lefebvre 1991, 6). This inclination to assume a measured space across the picture plane (whether in a perspectival painting, a photograph, or through a screen) as a given, as something that seems to perfectly describe Descartes's *res extensa*, shows the difficulties that surround the imbrication of perspective, representation, and space, and the perpetuation of landscape as a passively consumed visual object.

With these historical burdens is it possible to re-formulate landscape images in ways that can shape a different relationship with the landscape itself? How can we get around the problems of landscape and representation? In the following sections I want to re-evaluate the landscape image and its construction and function under Western norms. I will argue that heterotopia, as theorised by Foucault, can help to do this through "suspending," "neutralising," and "inverting" the assumptions of realism, of objective space, and representation that surround its regimes (Foucault 2008, 17). New imaginings are crucial for a different formulation of landscape and as Julian Reid shows, the "space-image" is crucial for shaping the imagination:

> [The] space-image . . . is not an image of space, nor simply a spatialized form of image, but a space in which images live, that is, the living space of the image. It is a space in which the imagination locates itself as well as an image in which the imagination sees itself, that is, the image of imagination.
>
> (Reid 2018, 45)

Reid defines the ship that Foucault invokes within his writing as a "space-image," a space that is also an image, that transports the imagination and in which it thrives (2018, 45). I suggest that the landscape image, rather than continuing as an image of rationally laid out "space," needs to become a "space-image" in order to establish a different imagining of landscape. It is through the "space-image" that landscape can be presented in non-perspectival ways that move it away from its association with representation and towards the persistent and shared space of materiality. The structures behind the image of landscape in Veronese's painting still shape ideas of landscape today, and our digitalised technologies are able to appropriate visual space in ways far more efficient than Renaissance perspective. The globalisation processes that drive the early twenty-first century are not remote from or irrelevant to this predicament, for the digital image stream propels them and reaffirms their ideological assumptions. New imaginings of what landscape might be are urgently required, imaginings that enable us to see it and our relationship with it, in ways adequate to the seemingly insurmountable problems of anthropocenic climate crisis and the planetary consequences of globalised capitalism.

Landscapes of difference

My practice as an artist engages with these ideas and asks whether pre-perspectival landscape backgrounds can be re-configured so that they critique this explicitly visual construction of landscape. Properties specific to the heterotopia, laid out by Foucault and elaborated on since in various ways, can be applied in order to enable a nuanced understanding of the landscape image, formulating it as a discursive object that calls into question previous dichotomies, that becomes an active agent in the way space and environment are imagined and therefore shaped. This process defines landscape as a co-subject by resisting the overtly optical effect produced by dominant (now digital, increasingly virtual) image forms and loosening the fixity of perspective's viewing subject/viewed object matrix. In doing so the image opens up to difference. Depicted fragments of wilderness landscape from fourteenth- and fifteenth-century Italian painting (that crucially offer no perspectival view) can return to the present as translated forms that challenge the assumptions of a virtualised, digital image world and propose an alternative to landscape as representation. Through their appropriation, these pre-modern landscape forms become reconfigured as radical propositions for our visual and spatial paradigms. Through the removal of all narrative elements (a process that also frees them from the specificities of their historical era) they reposition the idea of landscape from background towards the central focus of the image, no longer peripheral context, but a human-less subject in its own right whose material configurations are not subject to human narrative. In this way the rational language of representation is side-lined and these landscape forms leave behind the restrictive relationship they have with narrative – landscape as the "natural" setting for a "cultural" story. Their reformulation as twenty-first-century artworks unleashes their heterotopic force into a contemporary visuality underpinned by perspectival structures and shaped by global and digital capitalism.

This process of re-appropriation (analysed in the next section) exploits a practice of "othering," whereby the strangeness of these landscapes that operate outside perspectival landscape space and representational image-making disrupts the sameness of the digitalised landscape image in its globalised, circulatory form. Landscape elements such as those depicted within the fresco cycles painted by Giotto or Pietro Lorenzetti in the Basilica of San Francesco, Assisi, problematise the overtly visual, later history of the landscape genre, with its prioritisation of naturalistically rendered extensive space laid out in front of the viewer who imagines (as depicted by Veronese) that they are looking through a window and could possibly step out onto the landscape beyond.

Rather than understanding landscape as a kind of scenery in front of which the human subject appears, I utilise these forms in order to radicalise the idea of the background and to oppose the genre's presentation of landscape as a cohesive view that extends across the picture plane and that provides an image of "nature" in which the (human) subject is then contextualised through narrative. Any previous narrative or scenic elements are removed, opening up the purely visual to new temporal and material possibilities, and shifting the resonance of landscape from space to time. A displacement from landscape as image to landscape as fellow material body, even something fragile that requires care, occurs. A fragment is the opposite to an extensive view and my works prioritise materiality over visuality, demanding imaginative and affective contact with their appropriated landscape forms, disarranging flat visual representation in the process. In this way they also shift our received understanding of historical linearity, bringing the distant past into the contemporary; earlier forms rediscovered as "unfamiliar" acting on the present, positing a different potential for landscape. Indeed, as Foucault says of Borges's Chinese encyclopaedia, "the exotic charm of another system of thought, is the [realisation of the] limitation of our own, the stark impossibility of thinking *that*" (1989, xvi, italics in original).

If analysis from the field of art history primarily entails a process of looking and writing, the process of making (or "re-making") introduces a new form of contact with an earlier painting, which emphasises the material "other" of the image. And by exploiting the paradoxical qualities that are concealed within our habituated perspectival modes of viewing, attention is brought to a spatial present via a new material body that moves between then and now, critiquing the spectacularised capitalist subject that is always headed towards an idealised technological future. Following Foucault's definition of heterotopia, these works reconfigure the relationship between language and the world, and so help rethink landscape as a "co-subject," unfixing it from its structuring as "objective space." As Foucault argues:

> *Heterotopias* are disturbing, probably because they secretly undermine language, because they make it impossible to name this *and* that, because they shatter or tangle common names, because they destroy "syntax" in advance, and not only the syntax with which we construct sentences but also that less apparent syntax which causes words and things (next to and also opposite one another) to "hold together." . . . [H]eterotopias . . . desiccate speech, stop words in their tracks, contest the very possibility of grammar at its source; they dissolve our myths and sterilize the lyricism of our sentences.
>
> (Foucault 1989, xix, italics in original)

The next section explores this idea of the co-subject through a discussion of four artworks that actively exploit the inherent tensions within the "syntax" that underpins the landscape image including its relationship to narrative and to the human subject, to representation and to the technological appropriation of human vision associated with Western visual systems.

Landscape as co-subject

Landscape with a Spring (after Giotto) (Figure 7.2) explores what a landscape painting might mean if it contains no human narrative or subject. The painting shows a dry and featureless rocky outcrop in which, to the extreme bottom right, there is a small stream, a spring that emerges from a hole in the rocks. The painting copies from *The Miracle of the Spring*, part of the fresco cycle in the Upper Basilica in Assisi painted between 1297 and 1299 and attributed to Giotto (1266–1337). The frescoes show stories from the life of Saint Francis, as recounted by the Franciscan Prior General, Bonaventura. Here, Francis is travelling in the mountains with two monks and a peasant who (according to Bonaventura) has lent him his

Figure 7.2 Landscape with a Spring (after Giotto), Henrietta Simson, 2010, 48 × 60 cm, oil and pigment on gesso

donkey. The peasant is weak from exhaustion and in need of some water. Francis kneels down and prays, and a spring immediately appears which is subsequently never found again. The fresco thereby defines a place in which the heavenly has miraculously and momentarily broken through, revealing the already heterotopic formulation of this "landscape-place." According to Foucault, the "Middle Ages" were a "hierarchical ensemble of places" where a cosmology existed in which "the supercelestial places . . . opposed . . . the celestial, and the celestial place was in turn opposed to the terrestrial place. . . . It was this complete hierarchy, this opposition, this intersection of places that constituted what could very roughly be called medieval space" (Foucault 1989, 1). This hierarchy divided space, with each part understood in terms of its position within the whole. The world was shaped by symbols, whose language was hidden within it. Foucault's discussion of the heterotopia is linked to this larger project that explores the relation between language and "things," as exemplified in *The Order of Things*. He draws out the differing historical regimes of knowledge in Western thought, exploring ideas of representation and resemblance, and the persistence of language. Foucault sees that under the regime of representation (epitomised by Descartes's thinking "*I*") no resemblance is necessary between object and thought; a rational abstraction replaces the realms of symbolic correspondence that shaped the medieval world: "Representations are not rooted in a world that gives them meaning; they open of themselves on to a space that is their own, whose internal network gives rise to meaning" (1989, 86–87).

Language, the "word of God," shaped the medieval world through the idea of the "model"; it constituted the mystical origin of things, which could then reveal and bring forth its divine mysteries through symbols manifest in the physical world. In Giotto's narrative, the sudden appearance of the spring is confirmation of God's mysterious writing of the world. In this world of medieval resemblance, a deeper order (that of a divine mystery) is revealed through symbols. Conversely, in the world of representation, the world is "object" and this deeper order is revealed only in the thinking mind, which *interprets* the world and reflexively recognises itself and its relation to the divine. The "Word" has moved from the physical to the mental sphere and the landscape becomes extensive empty space bereft of miracles. In this world visual representation relies more on abstract mathematical qualities than it does on resemblance. However Foucault asserts that language persists in this visual, abstracted episteme of representation, although not in its previous existence, for "[i]t no longer appears hidden in the enigma of the mark" and not in future semiotic terms, "it has not yet appeared in the theory of signi-fication" (1989, 87). Accordingly, it reinforces the hierarchy and ideology of a scientific culture and its rational representations, a culture that emphasises the "naturalness" of its images, and landscape as objective space. The power of the heterotopia lies in its ability to break this apart because it breaks apart the order of language, interrupting the way it shapes things, undermining its relation to the spatial.

Giotto's fresco reveals the world as resemblance, divine mystery underscored in the sudden appearance of the spring. *Landscape with a Spring (after Giotto)* is of a different order – the title only suggesting that there was once a more signifi-cant event attached to the spring – while the narrative figures which orientate the miracle and which populate a world of resemblance are gone. As a heterotopic landscape it becomes unfixed from the previous medieval order of resemblance, and simultaneously presents an image of landscape that is not of the perspectival order of representation. The divine breaks through in an instant in Giotto's paint-ing; here by breaking apart the language of representation and the visual, it is possible to focus on the landscape as subject *in itself*.

Figure 7.3 Landscape without Stigmata (after Pietro), Henrietta Simson, 2008, 65 × 60 cm, oil and pigment on gesso

This absence of human narrative signals the lack of human subject. In *Landscape without Stigmata (after Pietro)* (Figure 7.3), Pietro Lorenzetti's (c.1280–1348) fresco of *St. Francis Receiving his Stigmata*, painted in c.1320 also in the San Francesco Basilica (Lower Church) in Assisi, is reconfigured into a small panel painting from which St. Francis and the other friar, the small chapel, the bridge, and the seraphic vision have all been removed. Only the landscape remains – a rocky wilderness, barren except for two trees that punctuate the skyline, the dark blue of which is covered with a muted earth-red wash. A ravine zigzags vertically down the centre of the painting, and a small stream flows down this, spilling out into the dark pool that fills the bottom of the picture plane, giving the barren mountains a watery base which acts as a visual barrier to the viewer. This painting highlights the vulnerability of the object (or landscape) exposed and laid bare by the lack of human narrative. This sense is reinforced by the brittle, fragile gesso surface on which the landscape is painted, which contributes to a feeling of pathos in the work, pointing to the need to care for the "other" – as object or landscape. The ravine, as "stigmata" suggests that the landscape has been "wounded," by a presence no longer visible. This points to a new ecological ethics, where the lack of subject or figure, the "bracketed" human (as defined by Jane Bennett) (2010, ix) reveals the significance of the object in itself.[2] The scene is replete with an entirety of matter, as if one is looking in on a distilled and distant world, human presence evident (but not privileged) not simply by dint of it being an artefact, but also by the landscape's *embodied* character. This painted landscape does not represent a "natural space" in terms of Cartesian logic or the perspectival tradition of landscape, but instead operates in terms of affective materiality and imagination. It is a landscape that carries its own histories, and the traces of human subjectivity are visible within these, but it does not employ perspectival space to situate and implicate the viewer within the scene; its focus is instead on the landscape in its own right, and the viewer is confronted with a landscape that no longer frames the human subject, but that has re-positioned itself beyond the former subject/object hierarchies that shape the Western landscape tradition.

Crucial to the definition of the pictorial object as defined through its affective materiality and not in terms of language (or human narrative), is the idea of distance.[3] And by identifying the landscape image as "distant" it becomes easier to see it as a body in its own right, rather than its characterisation as a view that is required to contextualise the subject through narrative. Distance here is emphasised as historical rather than spatial, landscape as the ultimately inaccessible "other." In an apparent aporia, redefining visuality and materiality in this way can foster a relationship between the human body and the landscape form that is founded on a shared embodiment. The process of removing the traces of subject matter so that the landscapes can stand as singular objects suggests affinity through the affective, a bodily rather than rationally based understanding, but this is a highly paradoxical empathy, these paintings being absolutely dissimilar in their emptiness.[4] There is no human subject depicted in them by way of narrative; they cease to be "scenes" or "scenery" in which human drama is enacted, and instead become replete and "distant." And as Timothy Morton says (not without irony): "To love extension . . . is to love the thingly quality of the other, in the ultimate, Cartesian sense: to respect what is truly other about the other" (2007, 179).

Morton's argument reveals how "Nature" – the Cartesian "extension" of matter and space through which landscape is defined – is a constructed and problematic category, and he challenges the subject/object dualist constructions of the Western philosophical tradition. But through "loving" the "other" (in this case the *res extensa* of Descartes's *cogito*), we begin to empathise with that which is truly not the human subject. By recognising the distance within this shared but separated embodiment, landscape overcomes its role as backdrop to human life. Landscape as visual experience and landscape as material object become equally distributed within this work.

Hills and Other Spaces (after Martini) (Figure 7.4) develops a critique of the landscape image by drawing on Foucault's discussion of language as the hidden heart of representation that shapes and controls the rational understanding of

Figure 7.4 Hills and Other Spaces (after Martini), Henrietta Simson, 2014, 64 × 49 cm, oil on gesso panel

images. It explores the tension between representational images and language. This tension, concealed and fundamental to the hierarchical formation of a sepa-rate human subjectivity, informs our perception of images as rationally drawn spatial structures. The painting again straddles two historical eras, but it also straddles different spatialities and in doing so reveals the assumptions of the perspective paradigm and how this maintains the visual order of subjects and objects. The discussion of the previous two works was focussed on how the land-scape can become a discursive object; here the focus is perspectival – how the assumptions of rational language and natural vision hide the fractured nature of the human subject (a fracturing rich in potentiality). The work is loosely drawn from Simone Martini's (1284–1344) equestrian portrait of *Guidoriccio da Fogli-ano* in the Sala del Consiglio in Siena's town hall, painted in c.1330. It consists of two opposing hills, bleak with no distinguishing features other than a small lake on the top of the flat-topped hill to the right. The rocky ground is pale and arid, the effect a combination of the smooth but irregular surface of the gesso ground and a layering of painted washes. The contrasting sky is a deep blue, implying a night scene. In the sky, slightly to the left and high up towards the top edge of the panel, are two black circles that suggest orbs or holes. As orbs, they form two dark stars, but the impossibility that the light in the painting that falls across the hills and lake emanates from them instantly alters the viewer's perception, and they instead become holes in the gesso surface. They oscillate, being part of the pictorial illusion (as orbs) then part of the structure of the panel (as holes). Either way they are interruptions and break up the representational order of the painting. By doing so, they force it away from illusion so that the viewer is confronted with their own viewing, and recognises the image's and their own materiality within a more fully described visuality. As De Cauter and Dehaene state in their discussion of the triadic notion of heterotopic space, "Other spaces are *alternative* spaces, *altered* spaces, and often also *alternating* spaces, in the sense that two different time-spaces come together and switch from one into the other" (2008, 93, italics in original).

Hills and Other Spaces is an "alternating" space that exploits the inherently unstable nature of the viewing subject (and by extension the perspectival image), uncanny in its construction around the presence and absence of sight.[5] The black circles in the painting act like the black holes of the pupils, two eyes viewing the image from an unconnected point, so that the act of viewing or looking is revealed to the viewer, but from the opposite side of what is viewed, uncannily reminding them that the unified self is an idealised impossibility, the fantasy presented by capitalist images. As "holes" in the image's surface, these eyes become blind spots that key absence – the absence of light and of sight, and of subjectivity. In order for globalised visuality to move beyond representation, these fallibilities need to be brought into continuous play and not disregarded in idealising constructions of vision and the self. And this is ultimately an appeal to the uncanny, to the unfamiliar that abides within the familiar as described by Freud, and used by recent writers such as Morton, to build new arguments for landscape and the human.[6]

Christine Boyer has recently discussed the problems of representation expounded by Foucault in his writing on the paintings of Velazquez and Manet, and recognises the presence of ideology in these "natural" images that are understood in terms of the mirror when she asks, "How is it that representation, an illusory image formed in the mirror, kills imagination and critical perspective?" (Boyer 2008, 70). This ideological construction of the painting as an image that passively and accurately *reflects* space as opposed to actively *constructing* it through language (an image that we "take as read" and assume to be "accurate") is exposed in Foucault's discussion of *Las Meninas*. Here he draws out the differences (and collusions) between the representational painting and the mirror, and between language and image and how these construct the viewing subject. As Boyer states,

> Recurring in Foucault's discourse on "other" spaces and on painting is the place of the spectator-subject, along with metaphors of the gaze and the mirror. In obscure and confused operations, Foucault is posing the problem of how the visible confronts the articulable, and how a counter-site is opened up.
> (2008, 61)

The visible and sayable, meshed in an uncomfortable complicity under the order of representation, when shown to be incompatible (in Foucault's argument through the operations of the heterotopic mirror) allow the viewer to see how this "natural" vision is shaped by rational language. In an instant they are able "to comprehend the imaginary experience of the gaze via its imaginary displacement" (Boyer 2008, 62). This is how the black circles in *Hills and Other Spaces* function. By detaching the image from perspectival norms, where the spectator is always implied, the latter is confronted with ambiguity. And it is within the space of uncertainty that the imagination can move, and that visual orders can reconfigure.

The processes of layering different spaces (so that that which is represented within the image is juxtaposed with the physical space of the viewer) and/or times (the historical past of the artwork and its reformulation within the present) are forms of spatial-temporal disruption, heterotopic actions that break apart the implied cohesive whole of the landscape image, replacing this with fragmented constructions that do not privilege the visual human subject. The strongly felt sense of embodiment within the medieval and Renaissance image is utilised so that these works can contest the present incarnation of the subject under capitalism, a virtual avatar, constructed through the latest (digital) perspective technologies. These govern and order visual space so that images become space *per se* – lived space conflated with representation – the former thus remaining inert and unconstructed, while the material reality that constructs the image is effaced (this effacement required if "natural vision" is to be simulated effectively). As a means to move beyond this, it is necessary to pull apart the psychological layering of the two-dimensional image's "screen" – its symbolic interface – so that this imbrication of visual, actual, and pictorial space can be unpicked. In Foucault's "third principle" he states that "[t]he heterotopia is capable of juxtaposing in a single real place several spaces, several sites that are in themselves incompatible" (Foucault 1986, 25). These paintings, "incompatible" in their non-representational, non-perspectival spaces act as "holes" in the main spatial milieu, and through a productive process of de-familiarisation, expose to ourselves our own sense of space, of landscape, inviting us to see anew.

Di Paolo Blue Wilderness (Figure 7.5) is a work that develops this idea of layering, exploring it via contrasting spatial and technological processes, combining sculptural forms, digital photographic manipulation, and pre-perspectival image space, and in doing so shows the technological elision of representational images and spatial experience. The modelled clay hills appear incongruous within the projected digital space and the technologically produced perspectival image is examined via this rearrangement, its paradoxical properties exposed. The work uses the wilderness landscape in which John the Baptist, depicted by Giovanni di Paolo (1403–1482) in 1454, roams, to defy the definitions imposed by the established terms of perspectival landscape or the textual de-construction of these, and instead focuses on an affective response to the material construction (rather than the removed or disregarded narrative forms) of the work. It draws out embodied connections and presents space in materially orientated ways rather than as that

Figure 7.5 Di Paolo Blue Wilderness, Henrietta Simson, 2018, 50 × 36 cm, digital image, oil on clay

which is rationally cohesive and understood in purely visual terms. This recasting of the landscape image as embodiment – as opposed to a space constructed through the perspectival – moves beyond these restrictive categories, towards the notion, set out by Dehaene and De Cauter, of the heterotopia as *play* (2008, 87–102). In their argument they draw out the duality of labour and action that Hannah Arendt describes as "the human condition," introducing a third space that incorporates the "([now] mostly secularized) sacred space" of cultural and playful activities that they see as the innately creative and "irreducible" aspects of life (De Cauter and Dehaene 2008, 95). They define the Greek theatre and other recreational spaces as part of this heterotopic intermediary "third sphere" that mediates between the public and the private through a sense of play. *Blue Wilderness* "plays" with the norms of Western visuality, the habituated spatial perceptions produced under the assumptions of representation, what a landscape should do or be in a neo-liberal world, so that the knotted relationship between the landscape image and the landscape as a recreational space is revealed.

As a photographic or perspectival image, landscape becomes the desired object – the space itself, a conundrum recognised by Hubert Damisch as the fascination that perspective painting holds for us. Christopher Wood describes it thus:

"We know the representation is not reality; and yet to a point we react to it as if it were real. Damisch calls this the double articulation of painting, representation and presence" (Wood 1995, 678). This "double articulation" emphasises the commodification of landscape under contemporary capitalism. It is sign and substitute simultaneously. It is a place of depicted recreation where one might go to escape from the travails of labour. Wood has discussed how these formulations are remarkably consistent throughout the history of landscape under capitalist visuality, where it is not only conceived as "parergonal" in terms of its function within the painting, but also in terms of work: "Recreation in Western pictorial culture . . . is meant to follow work and therefore stand outside it. It is, perhaps, a reward for work completed. But pleasure also prepares one to resume work by restoring or recreating the spirit" (Wood 1993, 55). This not only reflects contemporary attitudes to landscape, but its definition as recreational "parergon" is also found within Thomas Blount's Glossographia of 1670, in the period that popularised the genre of landscape painting in the West.[7] Landscape today is a place of tourism and recreation, a place to visit for the weekend or for a holiday. As an image it is tied up with this idea, a sign of leisure, adjunct to the capitalist economy. In this form it is invariably shown in romanticised and idealised terms, always clean, perhaps majestic, ultimately waiting there in the beyond to "refresh the spirit" or provide an "experience" ("the holiday of a lifetime"). These images work as excessive signs, as if the refreshment only needs to happen visually. Likewise Wood describes how the act of painting sixteenth-century landscapes was seen as respite enough for the painter, it was not necessary to actually go anywhere; the process of painting itself was sufficient (Wood 1993, 55). Landscape in this form is visually consumed. Indeed, as Nicholas Mirzoeff underlines, "[c]apital has commodified all aspects of everyday life including the human body and even the process of looking itself" (1999, 27).

Dehaene and De Cauter explore the heterotopia through the idea of the holiday, with its etymological reference to what is holy ("holi"-day) underscoring the importance of recreation. They state that "[h]eterotopia is perhaps more easily identified by its time than by its space. It is not simply a space but rather a time-space" (De Cauter and Dehaene 2008, 92). Although the "time-space" of the commodified landscape image is fleeting (it is the interstitial "time-space" of the screen-saver), more widely, the leisure space of landscape under capitalism frequently acts as the space in which this "holy" time occurs. It becomes a "time-space." This "time-space" also describes the function of the early wilderness landscape, the caves and arid mountains depicted in the paintings that shape my practice. Located outside habitual living space, the wilderness landscape operates within its own time, a liminal place of transformation where life and death exist in close proximity. As depicted within these early paintings, saints and holy men and women departed to these places, withdrawing from society (and from the body) to prepare the soul for entry into heaven. These alien landscapes, removed from habitual patterns of time and space, facilitated penance. *Blue Wilderness*, itself removed from habitual patterns of time and space with its non-corresponding elements, is a landscape that invites the imagination in, not for the distracted or passive enjoyment of the fleeting commodified image, but to be actively engaged in its suggestions and complexities.

Unfamiliar terrains

By appropriating the landscape forms of historical painting, through a process that layers and rearranges different temporal strata and in doing so breaks down the fixity of representational space, it is possible to access difference from within the dominant perspectival systems in a way that enriches our virtualised visuality. The process of removing the human subject – so central to these works – identifies how visual representation – the iconic – has operated in terms of the linguistic; how vision and narrative have been allied since Alberti's humanist concerns for perspective and the Cartesian annexing of the visual within rational thought. By directing attention towards the affective possibilities of the image that arise from its embodied forms rather than from its narrative elements, it is possible to instead explore an idea of a shared materiality that collapses the dualistic hierarchy of human subject/landscape object and that shifts the notion of distance from a "view" to a material background brought close, or a distant past brought present. This process interrupts the perspectival norms by which we habitually define images, and instead directs their appeal towards that which is affective and unfamiliar, accessing the uncanny and revealing the representational image's ultimate failure to represent its object. The works exist in a form of "inbetweenness," and in this way they stretch the dualisms that are held to be so restrictive, establishing new possibilities for landscape as a category. As Hilde Heynen suggests, "Pursuing the idea of heterotopia offers a productive strategy . . . because it introduces a third term in situations where strict dichotomies – such as public/private; urban/rural or local/global – no longer provide viable frameworks for analysis" (2008, 312). Utilising this "third term" within landscape and its representations opens up the dualistic problem of landscape/image, the heterotopia providing a useful methodology for moving beyond the impasse of the landscape image in its traditional forms. These landscapes move away from the dichotomies of nature/culture, subject/object primarily because they resist definitions set out by the genre of landscape painting and its perspectival structuring. They are not images of "natural" landscapes but cultural appropriations, and they are no longer the "object" through which the "subject" is constructed. As "outsiders" they escape the (perspectival, ideological) landscape way of seeing, and crucially, as parergonal forms they can be unfixed from their original time and can act on the present in ways that do not lock them into strict definitions as "previous" or "historical." This "in-betweeness" presents an inherent instability: they are not past nor strictly present, they are not nature and they do not establish an idea of landscape by using traditional landscape tropes. The works interrupt visual/spatial norms through this unbuckling of chronological syntax. This is a profoundly heterotopic process, one that reconfigures the habitual definitions of landscape in terms of human narrative, and uproots these unfamiliar background forms from their supportive roles within their original locations. The reintroduction of these landscape forms into this culture, causes a disturbance in capitalist space-time, creating a "heterochrony" – a "slice[] of time" – as Foucault calls it, that interrupts time's linear progression as something that we pass through (2008, 20). Instead time accumulates, so that we are removed from its habitual patterning as streamlined and linear. Indeed, these landscapes do not operate within the spatial norms of contemporary Western culture either, disrupting these by the addition of their accumulated time. Spatial configurations that are not constructed by a Cartesian representational system are brought forward through the 700 years or so that their landscape forms have endured. They slip between past and present, "space-images" that re-focus the imagination on the landscape itself, its accrued time and its material body that is not just an image.

In *Specters of Marx*, Jacques Derrida establishes an ethics of appropriation where historical forms are not so much taken possession of, but rather the complex operation between "ghost" and present purpose is acknowledged and the historical form allowed to "speak" (Derrida 1994). These "ghosts" offer forgotten alternatives and present new potentialities. Rather than blending difference into the hegemonic conditions of globalisation, the appropriated forms can instead destabilise the spatial-temporal norm. This action defines the works here as part of a critical strategy of remembering, working in counter to the "eternalised present" of the Internet, and its specifically dislocated and spectacularised form of subjectivity that shapes twenty-first-century capitalism. Rather than being made from adapted commodities or directly carving out new meanings for historical objects, the works exist within the terms of a translation or transcription. The appropriated historical forms bring their own language to notions of the contemporary image, whose digital and cyber context is assumed in a fixed technological trajectory. As Hal Foster states, "The deployment of the outmoded . . . can still query the totalistic assumptions of capitalist culture, never more grandiose than today" (2004, 16).

Foucault's project was always focussed, as Boyer suggests, on the "active engagement in liberation movements for prisoners, asylum inmates, student resistances – all targets of administrative power and all inspired by the dream of a radical subjective freedom" (2008, 63). A new ethics can be found within the space of the heterotopia, where these paintings become discursive objects that act as facilitators and extend this "subjective freedom" to the landscape itself. They show how the previous division of subject/object that enforces the carefully separated forms that constitute images of landscape are problematically rigid in a world where such division is no longer possible, not even at bedrock level. Exploring these older visual conventions in the context of digitally produced and disseminated images also locates Western visual culture within its technological history. Materiality and visuality are realigned and this has implications for our spatial imaginings, and for how subjectivity is shaped. These works contribute to a discourse that challenges the removal of space (as landscape here) to the abstract realm of representation, where its radical potentiality is reduced. The need to understand space in broader and more dynamic terms is urgent, and the landscape image – rather than reinforcing it as representation – can do this via the actions of the heterotopia.

Notes

1 Works such as *Vision and Painting* (1983) by Norman Bryson supersede previous definitions of the image as put forward by art historians such as Ernst Gombrich in *Art and
Illusion* (1977).
2 In *Vibrant Matter* (2010), Jane Bennett explores the question of the inherent agency
within matter in terms that remove the hierarchical binary between subjects and objects,
while not collapsing her thinking into a disempowering refusal of the human. In order to
do this she "brackets" the human and re-examines social processes, accommodating the
non-human components that exist within them.
3 For a discussion of images in terms of distance see Jean-Luc Nancy, "The Image – the
Distinct" in *The Ground of the Image* (2005). For the re-thinking of distance as a positive
and essential characteristic of landscape see John Wylie, "The Distant. Thinking toward
Renewed Senses of Landscape and Distance" in *Environment, Space, Place* (2017). I am
indebted to both arguments in shaping my own ideas of distance and landscape in my
practice.
4 In his introduction to Deleuze and Guattari's *A Thousand Plateaus*, Brian Massumi
defines affect as a "prepersonal intensity" (1987, xvii). He later explores it as related
(but prior) to "feelings" (personal) and "emotions" (social) in "The Autonomy of Affect"
(1995) and in *Parables for the Virtual: Movement, Affect, Sensation* (2002).
5 Much has been written about the uncanny nature of the gaze. See especially Jacques
Lacan, *The Four Fundamental Concepts of Psycho-analysis* (1998).
6 For the uncanny in terms of landscape, see Jean-Luc Nancy, "Uncanny Landscapes"
(2005); Tim Morton, *The Ecological Thought* (2010); and John Wylie, "The Distant"
(2017).
7 See Malcolm Andrews (1999, 30). See also Jacques Derrida (1987) for a discussion of
the parergon's importance in the work of art.

Works cited

Alberti, L. B. 2004. *On Painting*. Translated by Phaidon Press. London: Penguin.

Andrews, Malcolm. 1999. *Landscape and Western Art*. Oxford: Oxford University Press.

Baxandall, Michael. 1988. *Painting and Experience in Fifteenth Century Italy: A Primer in
the Social History of Pictorial Style*. Oxford: Oxford University Press.

Bennett, Jane. 2010. *Vibrant Matter, a Political Ecology of Things*. Durham and London:
Duke University Press.

Boyer, M. Christine. 2008. "The Many Mirrors of Foucault and Their Architectural Reflections." In *Heterotopia and the City: Public Space in a Postcivil Society*, edited by
Michiel Dehaene and Lieven De Cauter, 53–74. Abingdon: Routledge.

Bryson, Norman. 1983. *Vision and Painting: The Logic of the Gaze*. New Haven and London: Yale University Press.

De Cauter, Lieven, and Michiel Dehaene. 2008. "The Space of Play: Towards a General
Theory of Heterotopia." In *Heterotopia and the City: Public Space in a Postcivil Society*,
edited by Michiel Dehaene and Lieven De Cauter, 87–102. Abingdon: Routledge.

Deleuze, Gilles, and Felix Guattari. 1987. *A Thousand Plateaus: Capitalism and Schizo-
phrenia*. Translated by Brian Massumi. London and New York: Continuum.

Derrida, Jacques. 1994. *Spectres of Marx: The State of the Debt, the Work of Mourning, and
the New International*. Translated by Peggy Kamuf. New York and London: Routledge.

———. 1987. *The Truth in Painting*. Translated by Geoff Bennington and Ian McLeod.
Chicago: University of Chicago.

Foster, Hal. 2004. "An Archival Impulse." *October* 110: 3–22. Cambridge, MA: The MIT Press.

Foucault, Michel. 2008 [1967]. "Of Other Spaces." In *Heterotopia and the City: Public Space in a Postcivil Society*, translated and edited by Michiel Dehaene and Lieven De Cauter, 16–17. Abingdon: Routledge.

———. 1989. *The Order of Things*. Abingdon and New York: Routledge.

———. 1986. "Of Other Spaces." Translated by Jay Miskowiec, *Diacritics*16 (1): 22–7.

Gombrich, E. H. 1977. *Art and Illusion: A Study in the Psychology of Pictorial Representation*. London: Phaidon.

Heynen, Hilde. 2008. "Heterotopia Unfolded?" In *Heterotopia and the City: Public Space in a Postcivil Society*, edited by Michiel Dehaene and Lieven De Cauter, 311–23. Abingdon: Routledge.

Jay, Martin. 1993. *Downcast Eyes: The Denigration of Vision in Twentieth-Century French Thought*. Berkeley and London: University of California Press.

Lacan, Jacques. 1998. *The Four Fundamental Concepts of Psycho-analysis*. Translated by Alan Sheridan. London: Vintage.

Lefebvre, Henri. 1991. *The Production of Space*. Translated by Donald Nicholson-Smith. Oxford: Basil Blackwell.

Massey, Doreen B. 2005. *For Space*. London: Sage.

Massumi, Brian. 2002. *Parables for the Virtual: Movement, Affect, Sensation*. Durham: Duke University Press.

———. 1995. "The Autonomy of Affect." *Cultural Critique, "The Politics of Systems and Environments"* 31 (Part II): 83–109.

Mirzoeff, Nicholas. 1999. *An Introduction to Visual Culture*. 1st ed. London: Routledge.

Mitchell, W. J. T. 1986. *Iconology: Image, Text, Ideology*. Chicago and London: University of Chicago Press.

Morton, Timothy. 2010. *The Ecological Thought*. Cambridge, MA and London: Harvard University Press.

———. 2007. *Ecology Without Nature, Rethinking Environmental Aesthetics*. Cambridge, MA: Harvard University Press.

Nancy, Jean-Luc. 2005. *The Ground of the Image*. Translated by Jeff Fort. New York: Fordham University Press.

Reid, J. 2018. "The Living Space of the Image." In *Spaces of Crisis and Critique: Heterotopias Beyond Foucault*, edited by David Hancock, Anthony Faramelli, and Robert G. White, 39–56. London: Bloomsbury Academic.

Steyerl, Hito. 2011. "In Free Fall: A Thought Experiment on Vertical Perspective." *e-flux* journal (24) (April). www.e-flux.com/journal/24/67860/in-free-fall-a-thought-experiment-on-vertical-perspective/.

Wood, Christopher S. 1995. "Reviewed Works: 'The Origin of Perspective,' Hubert Damisch, 'Le Jugement de Pâris,' Hubert Damisch'." *The Art Bulletin* 77 (4): 677–82.

———. 1993. *Albrecht Altdorfer and the Origins of Landscape*. Chicago and London: University of Chicago Press, Reaktion Books.

Wylie, John. 2017. "The Distant: Thinking Toward Renewed Senses of Landscape and Distance." *Environment, Space, Place* 9 (1) (Spring): 2–20.

8   Of tourists and refugees

The global beach in the twenty-first century

Ursula Kluwick and Virginia Richter

A diverse patch of seaside: the beach in the twenty-first century

In summer 2018, the exhibition *The Great British Seaside* at the National Maritime Museum Greenwich brought together photographs of British beaches from the 1960s to the present. One of the most recent exhibits, a photograph by Martin Parr (2017, Figure 8.1) depicting the beach at Southend-on-Sea, on the north side of the Thames Estuary, showed a varied group of people entering the sea: two dark-skinned teenage boys, a fully dressed woman wearing a hijab, a man in a pair of Union Jack bathing trunks as well as some swimmers who are already fully immersed in the water (Parr 2018, 77).

In her introduction to the corresponding catalogue section of *The Great British Seaside*, Susie Parr comments on Southend's central beach area:

> Here, holidaymakers from north-east London and the Essex conurbations wander in vibrant family groups. Passing by, you catch snippets of conversation in a multitude of languages: Urdu, Yiddish, Arabic, Polish, Mandarin, Italian. Could this be the most concentrated diverse patch of seaside in the world?
>
> (2018, 76)

The photograph and comment precisely capture the mixed configuration of the beach in the twenty-first century, as a site where heterotopia and globalisation intersect. Easily accessible from London's eastern agglomeration, the Essex beaches form an integral part of metropolitan leisure culture. Despite its decline as a holiday destination from the 1960s – typical of British seaside resorts in the time of cheap international travel – Southend is still visited by about 7.5 million tourists per year (Southend Rising n.d.). Its seaside, including nine separate beaches, a pier built in 1830 and various leisure facilities, thus constitutes an "other place," a site of recreation distinct from everyday life, for the working population of the London metropolitan area (Southend-on-Sea Borough Council n.d.). The diversity of languages overheard on the beach, bearing testimony to the ethnic and cultural – but perhaps not national, as many of these will be British citizens – diversity of its

Figure 8.1 Keystone/Magnum Photos/Martin Parr

visitors, is the result of what could be termed slow globalisation: Yiddish, Italian, Polish, Urdu, Arabic and Mandarin speak of the successive waves of immigration to the United Kingdom since the late nineteenth century. As many photographs shown in *The Great British Seaside* exhibition highlight, the beaches previously exclusive to an all-white English working class, such as Scarborough, Eastbourne and Blackpool, have now become shared spaces, visited by a population that reflects the heterogeneity of the British populace.[1]

The beach is not listed among Michel Foucault's famously motley examples of modern heterotopias (1986). With the help of recent theoretical reformulations of the concept, however, it can be argued that the beach constitutes a heterotopia *par excellence* in the era of globalisation. Drawing on classical sources such as Hannah Arendt's *The Human Condition* (1998) and Aristotle's *Politics*, Lieven De Cauter and Michiel Dehaene propose a tripartite division of the *polis* into private space (*oikos*), public space (*agora*) and "other spaces" which are dedicated neither to (private) economic pursuits nor to (public) political action, but to practices pursued for their own sake and associated with the sacred (*hiéran*). In the classical Greek city, such activities include sports and drama, and the third or hieratic space of the *polis* encompasses the theatre, the stadium and the gymnasium (De Cauter and Dehaene 2008, 90). As a space conceived in opposition to the political as well as the economical, it constitutes a heterotopia in Foucault's sense, a site that is explicitly marked as other (De Cauter and Dehaene 2008, 90) and that functions as a space of potential change. Importantly, this third space that is neither fully private nor fully public is associated with "what we commonly describe as the 'cultural sphere': the space of religion, arts, sports and leisure" (De Cauter and Dehaene 2008, 91). This triadic conceptualisation of urban space not only escapes the binary division into a public and a private sphere, but opens up the possibility of a dynamic third space in which oppositions such as "exclusive versus inclusive, kinship versus citizenship, hidden versus open, private property versus public domain . . . are reshuffled and readjusted" (De Cauter and Dehaene 2008, 91). While taking theories of the classical *polis* as their point of departure, De Cauter and Dehaene's revision of Foucault's concept of other spaces aims at making heterotopia available as an analytical tool for "the contemporary urban condition" (6). In the following, we show that heterotopia is equally central for a space that has been traditionally seen as an antithesis to urban life, the beach.

Crucially for a reflection on the beach as heterotopia, hieratic space is non-instrumental, not associated with gain or fame, but with play, an activity that is exempt from economic and political considerations. Practices typically pursued on the beach are playful in this sense: sunbathing, daydreaming, swimming, playing ball games, and perhaps most emblematically, building sandcastles. According to Rob Shields, the beach constitutes an "alternative geography" – another word for heterotopia, perhaps – because it is linked to "socially marginal activities" (1991, 4), activities that are precisely neither economical nor political. Building sandcastles belongs to "a specific ensemble of practices" which determine what are "not only appropriate but also *natural* attitudes and behaviours for a beach" (Shields 1991, 60). The most ephemeral of pursuits, as the sandcastle will be washed away by the next incoming tide, this practice points to the different temporal order of the beach, linking it to childhood as well as to lunar chronology which determines the rhythm of the tides. In fact, De Cauter and Dehaene stress the temporal dimension of heterotopia – as a time of festivities or carnival – and link it to holidays, an "other time" as well as an "other space":

> The English word *holiday* has kept this reference to the "holy" origin of free time, rest and repose. Similarly to the way in which heterotopia interrupts the continuity of space, the holiday interrupts the continuity of time. Holidays, being extraordinary as opposed to the mundane, ordinary character of the everyday, are the permanent makers of the discontinuous moments on the calendar, pacing the continuous flow of everyday experience.
>
> (De Cauter and Dehaene 2008, 92)

Since the establishment of regular paid holidays as a fixed part of industrial production, the seaside resort has become a favourite destination for this "time away" from everyday working life.[2] The beach correlates with De Cauter and Dehaene's definition of heterotopia because it entangles oppositions dividing public and private space: as a site (usually) outside the city but easily accessible from it, it is situated between nature and culture; it is a space dedicated to leisure and relaxation, but dependent on the work done by the vacationer throughout the year, as well as on the work of local beach attendants to maintain the cleanliness and infrastructure that keep the beach serviceable for recreation. The accessibility of beaches varies greatly, ranging from enclosures belonging to hotels and holiday resorts, and reserved for their clientele, to communal beaches that can be visited for free. However, in every instance the beach is a cross between public space – as a site generally shared with others, where bodies are put on display to an even greater degree than on the *agora* of the city – and private space, because every visitor delimits his or her own spot by putting up wind screens, beach chairs, parasols or by placing a towel on the sand. Most importantly, the beach is a heterotopia in so far as it functions as "the space-time in which normality is suspended in order to give a place to 'the rest'" (De Cauter and Dehaene 2008, 96) – in the double sense of the word, as rest from work and as the rest that is usually excluded from everyday life: the enjoyment of childish play, the cultivation of the body and the suspension of cultural norms, most distinctly symbolised by the partial or complete shedding of clothes.

One reason why the beach to date has escaped much consideration as a heterotopia lies perhaps in the urban conceptualisation of other spaces.[3] From Foucault to Edward Soja's notion of "thirdspace" (1989, 1991) and the contributions to De Cauter and Dehaene's *Heterotopia and the City*, heterotopia has been discussed as a part of the *polis*, whereas the beach, at least in the collective imaginary produced by postcards, advertisements and touristic websites, lies outside, at the edge of civilisation. However, we want to argue that the relationship between the city and the beach is topographically and conceptually complex. The beach as we discuss it here – neither as private property nor as a remote wilderness – both is dependent on exchange with an urban centre and solicits the infrastructure that enables this exchange.[4] As the example of Southend-on-Sea discussed earlier shows, littoral recreational areas are either close to metropolitan areas, or fairly easily accessible by train, road, boat or airplane. In a different sense, today even far-away beaches have become part of the Global Village, available for viewing and consumption via Google, booking apps and social media. The increasing popularity of urban beaches, on the banks of the Seine in Paris or the river Spree in Berlin, further indicates why the beach, while functioning as an other space to economic and political space, is not starkly antithetic to the city.

Parr's photograph of Southend-on-Sea shows an ethnically and culturally mixed group of bathers. One could interpret this as a celebratory picture of "multicultural Britain"; however, the photograph also discloses potential tension. Into what kind of relationship could the man sporting the Union Jack on his bathing trunks enter with the Muslim woman? Are they, by their different attire, demarcating exclusive cultural spaces around their bodies, or are they, by their proximity and an exchange of smiles and glances, creating a shared space, a zone of mutual acceptance? Precisely because it is a site of unrestricted mixing, the beach denotes both inclusion and the contestation of space. If the beach is potentially open to everybody – in practice, only to those who can afford the trip and time away from work – it also has a long history of exclusion and even segregation.[5]

As Marco Cenzatti argues, one of the hallmarks of heterotopia in contemporary, post-Fordist society is the disintegration of public space, as a result of more decentralised, flexible sites of production, markets, social roles and power structures (2008, 77–78). Third spaces are now constituted through the interactions of many different social groups, rather than by top-down acts of normative institutions. In consequence, to Foucault's notion of the modern "heterotopia of deviation" (spaces assigned to those who more or less permanently do not fit into society, such as the prison, the mental asylum or the retirement home) another dimension should be added, the "heterotopia of difference" (Foucault 1986, 25; Cenzatti 2008, 76). As Cenzatti argues, heterotopias of deviation are still present, but another layer can be added. He writes: "Heterotopias of difference are still places in which irreconcilable spaces coexist, but what constitutes irreconcilability is constantly contested and changing. As these heterotopias fluctuate between contradiction and acceptance, their physical expression equally fluctuates between invisibility and recognition" (Cenzatti 2008, 79). Heterotopias of difference are thus flexible in terms of time and space: they can arise spontaneously, migrate from place to place, and assume different meanings across a given timespan (for instance, in the day and at night, during the working week and on the weekend).

This flux is a crucial characteristic to be added to our conception of the beach as heterotopia. Just as littoral terrain itself is marked by fluctuation and instability, the social space of the beach is constantly contested by various agents who visit, work on and profit from the beach: local inhabitants, tourists, developers, the tourist industry, homeless people who stay overnight, hawkers selling souvenirs to the vacationers, refugees who illegally come ashore, border patrols and the police. Some of these groups share interests with others, but often, interests irreconcilably clash. The beach can function as a hieratic space of play for vacationers or as a hieratic space of sanctuary for refugees after a dangerous crossing, but hardly both at the same time.

It is our aim to analyse the beach as such a precarious, contested and multidimensional heterotopia in literary texts. Specifically, we want to look at the differences of experience among those for whom the beach functions as a recreational heterotopia, in other words, wealthy tourists, and those for whom it is a work place, a site of displacement and violence or, in the case of refugees, a site of arrival and a holding area. Contemporary fiction is prolific in addressing the question of global inequality and the often lethal conflicts that follow from it. For novels engaged in this way, the beach serves frequently as a main setting, not least because the overwhelmingly positive associations the beach has for Western readers put the depicted disasters into sharp relief. Already Alex Garland's bestselling novel *The Beach* (1996), describing a British dropout's quest for a perfect other space, showed the conflicts between a community of travellers living illegally in a Thai nature preserve and local drug dealers. In a more recent example, Nicole Denise-Benn describes the way luxurious holiday resorts encroach on the lives of local villagers in Jamaica, transforming the beaches that used to belong to the community into closed areas reserved for the paying guests:

> There is a sign that reads NO TRESPASSING on the beach right where Thandi used to play as a child, which was once an extension of River Bank. The

hotels are building along the coastlines. Slowly but surely they are coming, like a dark sea. Little Bay, which used to be two towns over from River Bank, was the first to go. Just five years ago the people of Little Bay left in droves, forced out of their homes and into the streets.

(Denise-Benn 2017, 120)

Rather than bringing prosperity to the locals, the growth of the tourist industry is described as the rising of a "dark sea," a slow but persistent deluge that will deprive the inhabitants of River Bank from what used to be, not so long ago, their own heterotopia of play.

The non-instrumental or even anti-instrumental designation of other spaces stressed by De Cauter and Dehaene suggests that heterotopia has an inherently critical or liberating quality in an otherwise fully economised world; they even posit heterotopia as the antithesis to non-place (Augé 1995) and the concentration camp (Agamben 1998; De Cauter and Dehaene 2008, 5). However, the beach can serve as a good example to problematise such optimism. As we will show in the following case study of Chris Cleave's *The Other Hand* (first published 2008), in the era of globalisation the paradisiacal constellation of the beach can quickly segue into the one or the other: while the beach is the ultimate get-away for wealthy Western tourists, it can literally become a dead end for the poor and the subaltern of this world. Neither paradise nor sanctuary, for refugees the beach can indeed come to signify a state of exception in which they are deprived of their human rights, or even lose their lives.

"There was a beach next to the war": littoral heterotopia and state of exception in Chris Cleave's *The Other Hand*

Chris Cleave's novel *The Other Hand* (published as *Little Bee* in the United States and Canada) engages with questions of illegal immigration and asylum, and with the surprising entanglements between lives in the Global North and South that a globalised world creates. The novel tells the story of Little Bee, a young Nigerian who has fled to England because her village was destroyed and her family killed during the Nigerian oil war. She stays with Sarah, an English journalist whom she met on a Nigerian beach on which Sarah holidayed with her husband Andrew. On the beach, Sarah and Andrew were drawn into the conflict as witnesses of the oil company's killers' violent pursuit of Little Bee and her sister Kindness, and Sarah lost her finger in the attempt to save the two girls from death. When Little Bee arrives in England, Andrew commits suicide because he feels guilty for what happened on the beach. Little Bee stays with Sarah until she is arrested and deported. The novel ends on the same Nigerian beach, where Little Bee is arrested by Nigerian soldiers.

The representation of the beach in *The Other Hand* is intricately connected to the novel's use of narrative perspective. Little Bee and Sarah function as auto-diegetic narrators who tell their stories in alternate chapters. This "dual first-person narrative structure" (Perfect 2014, 158) shapes the manner in which "what happened on the beach" (Cleave 2016, 141) is related in Chapters 4 and 5, in which Sarah teases the background of their shared littoral experience out of Little Bee. Their traumatic beach experience is thus revealed in a flashback, first by Sarah, who mixes Little Bee's account with her own memories, and subsequently by Little Bee in her own words. This manner of narration results in a combination of internal and external perspectives that helps to emphasise the way in which conceptions of the beach clash. Furthermore, through the juxtaposition of these two women's narratives, British policy towards refugees is challenged as personal experiences of persecution, sacrifice and violence reveal the fault lines in international conceptualisations of sanctuary and in Britain's self-image as a "safe haven," as quoted from a 2005 UK Home Office report in the epigraph to the novel.

The Other Hand contains many heterotopian settings. It begins with Little Bee's accidental release from a detention centre in Essex, and hence from a place that encapsulates some of the traits of a heterotopia of deviation in the Foucauldian sense. The heterotopia of deviation re-appears towards the end of the novel, when Little Bee is caught and arrested, and imprisoned in various holding cells across London, among them at Heathrow Airport. She is subsequently deported by plane, a heterotopia of difference in which various irreconcilable meanings (the plane as a means of transport taking tourists to their holiday destination or businesspeople to work, as an extended prison for Little Bee and as a space of work for the guard who accompanies her) coexist. The most important heterotopia in the novel, however, is the beach, specifically Ibeno Beach in Nigeria. It constitutes the setting for the brief but life-altering first encounter between Little Bee and Sarah, and is thus at the core of the entire action in the novel. It is here that the heterotopia of play collides with different meanings of heterotopian space-time, shaped by the expectations of international tourists and by economic imperatives (specifically, the presence of oil and its importance for Nigeria in a global economy), respectively. This results in an eruption of unspeakable violence which also implicates the beach as a site of globalisation as both its function for tourism and its relation to natural resources point to its significance as part of a global economic network.

Sarah and Andrew go to Nigeria because they need "a holiday" to save their troubled relationship. "Maybe we need a change of scenery. A fresh start" (Cleave 2016, 238), Sarah claims, activating precisely the idea of heterotopia as a space-time devoted to "the rest" (De Cauter and Dehaene 2008, 96). Her hope that a temporary break away from their normal surroundings will permanently revitalise their marriage points to an excessive faith in the benefits of heterotopia as separate from and yet linked to normality. Activating a conventional beach fantasy of surf and sun, she believes that she knows precisely what to expect: "It's just a beach holiday. Come on, how bad can it be?" (Cleave 2016, 239).

As it turns out, it is beyond "bad," because on the beach, Andrew and Sarah find themselves not in the expected paradise, but in the midst of an oil war: "The tourist board . . . noted that Ibeno Beach was an 'adventurous destination.' Actually, at the time we went, it was a cataclysm with borders" (Cleave 2016, 143). Sarah repeatedly blames herself for not having been aware of the conflict, but as the novel makes clear, her individual ignorance is part of a wider global understanding: "The struggle was brief, confused, and scarcely reported. The British and Nigerian governments both deny to this day that it even took place" (Cleave 2016, 141). When Little Bee asks for asylum in Britain, she is told that "Nigeria is a safe country" (Cleave 2016, 196). Ironically, this general lack of awareness is part of the reason for which Little Bee is in danger: "The sisters had seen what had been done to their village. There weren't supposed to be any survivors to tell the story" (Cleave 2016, 145). The violent destruction of human settlements for the purpose of the creation of oil fields is not supposed to be acknowledged on the international stage or global access to Nigerian oil might be jeopardised.

The manner in which Sarah begins her narrative about their Nigerian holiday aptly expresses the clash between conflicting configurations of the beach: "That season in Nigeria, there was an oil war" (Cleave 2016, 141). The term "season" here is taken from tourism brochures and as such it is incompatible with references to violence and war. Their common mention here points to the inappropriateness of Sarah's touristic expectations, and the scene on the beach takes much of its force from the manner in which her and Andrew's conception of it as vacationscape is brought up short. The heterotopia of play which they have come to enjoy rapidly transforms into a thanatoscape as the two tourists walk out of their protected hotel compound right into a state of exception.

Giorgio Agamben theorises the state of exception as a condition designed as "a provisional and exceptional measure" (2005, 2) by which "sovereign powers suspend the constituted law" (Colebrook and Maxwell 2016, 7). As such, however, it is itself part of the "heart of law" (Colebrook and Maxwell 2016, 48), and a "technique of government" that "tends increasingly to appear as the dominant paradigm of government in contemporary politics" (Agamben 2005, 2). The state of exception is implemented in the form of different institutions, such as the detention centre or the camp, in which "individuals can be subject to various forms of violence without legal consequence on territory that is outside the normal juridical order" (Owens 2009, 572). Through this legally enabled legal vacuum, human subjects are reduced to what Agamben calls "bare life," the form of "simple natural life" which he derives from the Greek concept of *zoē* and which is "excluded from the *polis*" (1998, 2). These subjects are deprived of political agency and any rights of citizenship. For Agamben, it is in the figure of the refugee that bare life "fully enters into the structure of the state" (1998, 127), since "the so-called sacred and inalienable rights of man show themselves to lack every protection and reality at the moment in which they can no longer take the form of rights belonging to citizens" (1998, 126).

In *The Other Hand*, Little Bee clearly belongs to the realm of bare life: as a refugee and illegal immigrant, she has no rights of citizenship and no protection, even though her life is acutely threatened. However, it is not in an institutionalised context that the consequences of her status are most striking in the novel, but on the beach, where the state of exception making possible her reduction to bare life becomes all the more unsettling through its juxtaposition with the heterotopia of play.

In the scene taking place on Ibeno Beach, heterotopia and state of exception initially coexist. The beach is established as vacationscape through the manner in which Andrew and Sarah behave before they meet Little Bee and her sister. The couple are on holiday and Sarah in particular treats their littoral surroundings as a heterotopia of play that fits into the expected vacation paradigm. However, through the manner of narration, particularly through the alternating perspectives which we have already described, this conception is challenged from the beginning. The retrospective narration of events starts with the two sisters observing Sarah and Andrew from the edge of the beach, all the while conscious of the threat approaching from inside the jungle. Whereas readers, therefore, are immediately aware of the potential danger, Sarah only notices the paradisiacal aspects of the beach: the "sunrise," the "unbelievably peaceful" atmosphere and the beauty of "sea," "jungle" and "palms" (Cleave 2016, 146). It is Andrew who notices the treacherousness of this vision:

> "I'm still a bit scared, frankly. We should go back inside the hotel compound."
> The white woman smiled. "Compounds are made for stepping outside."
> (Cleave 2016, 146–47)

For Sarah, "stepping outside" seems not only easy here but also part of natural behaviour belonging to the set of practices that are normal on the beach (Shields 1991, 60). Yet her vision is caused by ignorance, a fatal misreading of this specific beach. For her, the compound is an integral part of the beach because its existence satisfies her desire for transgression, which this holiday is supposed to compensate: she promises to give up her affair, Andrew's discovery of which provided the immediate motive for their holiday, as they decided to try to save their relationship by escaping their everyday life. Andrew is closer to recognising the actual significance of the compound on Ibeno Beach. Its presence shows that an architectural structure had to be erected not only to keep the idea of vacationscape intact for tourists, but to protect them from real physical harm. The heterotopia of play, in this instance, does not stretch beyond the closely guarded borders of the compound and it does not include the beach.

Because they embrace (albeit somewhat reluctantly, in Andrew's case) the illusion of unlimited freedom constructed within the heterotopia of play, Sarah and Andrew literally stumble from heterotopia into the realm of bare life, without really realising at which point the one shifts into the other. Ironically, it is precisely at the moment when vacationscape comes openly under threat that Andrew decides to accept the idea of the beach as a well-deserved heterotopia of play. This sudden change of attitude can be seen as an expression of his sense of his own rights, of which heterotopian freedom forms an integral part. As a guard tries to force them back into the safety of the compound, the couple persist in reading his actions in the light of vacationscape. They dismiss the man as "that doofus of a guard" whose presence signals their inability to "do [their] own thing even for one minute" in the regulated environment of the holiday resort (Cleave 2016, 148).

Andrew, a celebrated political columnist for a British newspaper, approaches the situation in terms of stereotypical conceptions of the interaction between tourists and locals, offering the guard money to leave them alone and disregarding the two Nigerian girls' eventual appeal for help "as a classic Nigeria scam" (Cleave 2016, 152).

Even when faced with the concrete threat posed by the oil company's killer troupe, Andrew and Sarah initially refuse to accept the fact that their privileged knowledge of the world's ways no longer holds, trying to assert their position as authoritative and financially potent tourists from the First World. Sarah later refers to this as a form of "intellectual jet lag" (Cleave 2016, 152) which prevents them from mentally processing the seriousness of the situation. When threatened by the killer troupe, Andrew and Sarah fall back on their earlier strategy against harassment and offer the leader of the group money. His reaction to this offer, however, confronts them with the haphazard logic of the state of exception, in which rationality and savagery are not as clearly distributed as they suppose. The mercenaries' leader surprises Sarah with his knowledge of her home town Kingston-upon-Thames, and with information about his educational status: "'I know where Kingston is,' he said. 'I studied mechanical engineering there'" (Cleave 2016, 158). Having claimed potential membership of Western civilisation, he then spectacularly violates its rules by cutting off the guard's head.

Perfect reads this scene as the clearest expression of the novel's "condemnation of the way in which Britain administers immigration and asylum" (2014, 166), arguing that the fact that this killer was "allowed into Britain to study" shows that access to Britain depends on "economic capital" rather than humanitarian principles (165). However, we believe that the situation is more complex than this. The mercenaries' leader is certainly "horrifying," and his actions are "frightening" in their radical randomness and cruelty (Perfect 2014, 165). But as he repeatedly emphasises, he is "not a savage" and he does not "want to kill these girls" (Cleave 2016, 163). Little Bee later reveals that the leader did not take part in Kindness's subsequent day-long rape and torture but "was far off from his men" (Cleave 2016, 187), eventually swimming out into the ocean, where he disappears from her sight, probably having committed suicide (Cleave 2016, 189–90). The fact that he studied mechanical engineering at Kingston College perhaps points to his potential in a context different from the one in which he finds himself in his home country. It certainly tells us much about the complex continuities between colonialism and globalisation. "You're crazy," says Sarah to the killer when he calls them responsible for the guard's death, but his reply points to a different version of the story: "'I live here,' he said. 'You were crazy to come'" (Cleave 2016, 159–60).

Andrew and Sarah's ignorance of the oil war shows how much is at stake in the global economy which this stretch of sand between the sea and the jungle has suddenly come to represent. Ibeno Beach is bounded by a river "iridescent with oil" and "bloated with the corpses of oil workers" and a "malarial jungle" in which villages are burnt down and their inhabitants murdered to make room for oil fields (Cleave 2016, 143). And although Sarah loses a finger in this confrontation with the oil company's men, Kindness and the guard lose their lives. In this littoral state of exception, juridical laws might have been suspended, but economic laws remain firmly in place. The two sisters and the guard are economically insignificant; Andrew and Sarah, by contrast, are protected by the global economic power which their nationalities represent: two missing European tourists are likely to pose a problem for international relations. Even as they inadvertently walk into the state of exception, therefore, this state never fully engulfs them, and they are never really reduced to bare life. The economic power which they represent protects them. Thus the fact that they have been given a freebie by the state tourist board, and the manner in which they are kept ignorant of the "cataclysm" around them in order not to jeopardise tourism (Cleave 2016, 143), represents a side of the global economy in which the littoral heterotopian fantasy is not non-instrumental in De Cauter and Dehaene's sense, but is, in fact, itself a lucrative commodity. It is thus fitting that the tiny heteroropian bubble of Ibeno Beach bursts spectacularly when it reconnects with political reality in the form of the two persecuted girls who in their plight represent bare life.

The heterotopia of play is not the only bubble that bursts as it collides with the beach as thanatoscape in the state of exception. When Andrew refuses to cut off his finger in exchange for the two Nigerian girls' lives, claiming that the men are "just going to kill them whatever" (Cleave 2016, 164), he is, of course, right in pointing out that the suspension of law in the state of exception makes the outcome of any action unpredictable. In the novel, however, his refusal to give his finger for the lives of the girls is read (by Andrew himself above all) as a failure of humanity and, specifically, a failure to bring into practice the high ideals which he preaches in his weekly column. As Andrew stands on the beach, unable to say whether he wants "to save these girls" at possible cost to himself, Sarah remembers the final paragraph of his last column: "*We are a self-interested society. How will our children learn to put others before themselves, if we do not?*" (Cleave 2016, 161, emphasis in text). Faced with the existential chaos of the state of exception, Andrew is unable to conform to the standard implied in his own rhetorical question, and this failure will eventually lead him to suicide. Sarah, by contrast, complies with the rules which the troupe leader has spontaneously invented and cuts off her finger, paying with her flesh for the girls' lives. Surprised by her action, the leader feels compelled to follow his own arbitrarily imposed rules. He allows her sacrifice to count, though only for one life, turning Kindness into the price to be paid for Andrew's refusal to play his game. While Andrew and Sarah initially cling to the rules of vacationscape, the troupe leader thus superimposes a different set of rules onto littoral space, forcing the couple into a game that contrasts sharply with the imagined benignity of the heterotopia of play which they want to embrace.

Sarah's actions save Little Bee, but she can only preserve the girl's bare life; she cannot reinvest her with the rights of citizenships. A mere few weeks after her release from the detention centre, Little Bee is arrested and deported. Sarah returns to Nigeria with her, intent on protecting Little Bee with her presence and money. But on Ibeno Beach, soldiers catch up with them. Curiously, however, and even though it is unlikely that Little Bee will survive this second encounter with Nigerian soldiers, the novel ends with a somewhat soppy reaffirmation of the heterotopia of play, a multicultural celebration of Nigerian children playing with Sarah's white, blond son Charlie in the surf. As Little Bee looks her goodbye at Charlie, she believes – rather puzzlingly, given that the soldiers just attempted to shoot him – that "he would be free now." Her sentence, however, continues: "even if I would not" (Cleave 2016, 264). Hence the reaffirmation of heterotopia as liberating is doubly challenged here. First, it remains ideologically stale since it rests on Little Bee's rather too exuberantly embraced self-sacrifice. Second, the insipidness of the final littoral image makes the ending aesthetically dubious. There is no room for bare life within this sudden and improbably joyful splashing "in the sparkling foam of the waves" (Cleave 2016, 374).

Conclusion, or what to wear to a war?

As our discussion of *The Other Hand* suggests, the beach is a multidimensional terrain in which irreconcilable meanings come together to create a heterotopia of difference. These meanings are generated through practices on different levels, such as the division of space into safe and unsafe, monitored and free areas – in Cleave's novel, the hotel compound vs. the beach – and what Michel de Certeau conceived as tactics, of everyday, often trivial and unpredictable acts (1988, 29–30), adopted by individual visitors. Such tactics are often tied to sartorial choices, as we saw in Martin Parr's photograph discussed in the beginning. Acts such as putting on bathing trunks with a Union Jack or fully covering one's body and head claim the public space of the beach in different ways. The close proximity of such distinct attitudes can be interpreted, variously, as a happy multicultural and multiracial mixing, realising the potential of the beach as a heterotopia of play, or as an exclusionary and even aggressive contest about the legitimate use and cultural significance of the beach.

We find similarly diverging readings of the beach also in contemporary fiction, where the semantics of the beach can segue or dramatically flip over into its very opposite. In Alex Garland's *The Beach* the secluded paradise of a community of dropouts suddenly turns into a scene of massacre between different factions within the group. Similarly, as we have seen, Sarah and Andrew's dream holiday destination in Cleave's novel *The Other Hand* mutates into a site of horrendous violence. Here again, not only strategic divisions of and the protagonists' movement across space are decisive, but also minor tactical decisions such as the choice of clothes. As so often in the history of beach holidays, the woman's body and the question of appropriate beachwear stand at the centre of spatial contestation.

At the beginning of their confrontation on Ibeno Beach, the mercenaries' leader tears off Sarah's beach wrap. His motivation remains vague; it might be sexual, it might be the desire to intimidate, or to humiliate. Sarah cannot interpret his action because she cannot identify the logic according to which he operates:

> He made a sound, an involuntary moan which seemed to surprise him – his eyes went wide – and he tore off my beach wrap. He looked down at the pale lilac material in his hands, curiously, and seemed to be wondering how it had got there.
>
> (Cleave 2016, 156)

Significantly, this "wondering" is something that unites Sarah and this man at this moment. On Ibeno Beach, Sarah's beachwear comes to express the frightening absurdity of the sudden collision between incompatible meanings of littoral space. Dressed according to the rules of the heterotopia of play, Sarah experiences the shock of the sudden transformation of holiday into nightmare as a feeling of inappropriateness, not only of their very presence as tourists, but of her attire:

> I was wearing a very small green bikini. I will say that again, and maybe I will begin to understand it myself. In the contested delta area of an African country in the middle of a three-way oil war, because there was a beach next to the war . . . I was wearing a very small green bandeau bikini from Hermès.
>
> (Cleave 2016, 156)

The bikini here becomes the focus of Sarah's bewildered inability to reconcile her conception of the beach as vacationscape with the state of exception. Even later, this detail continues to haunt her as an expression of the utter incongruity of expected and actual beach experience: "It all seemed suddenly insane," she thinks after Andrew's funeral, "like wearing a little green bikini to an African war" (Cleave 2016, 288). This phrase, "wearing a . . . bikini to . . . war" conveys the incommensurateness of and, at the same time, the intricate connections between the local and global trajectories that have brought these characters together on this specific beach. The naming of a particular designer, the global luxury brand Hermès, further reinforces Sarah's status as a highly privileged tourist whose sense of invulnerability is pulled up short by the mercenary's forcible denuding of her body.

To return to De Cauter and Dehaene's reconceptualisation of heterotopia discussed in the beginning, in the contemporary global *polis* the beach functions only to a certain degree as a third space that is exempt from economic and political action. There is no such thing as a freebie, a holiday untouched by global processes such as conflict over resources (not only oil, but also sand is a valuable resource), the slow violence of global warming and pollution, and the unequal distribution of and access to wealth, security and human rights. Nevertheless, because so many trajectories intersect on the beach, it does function as a dynamic space in which social change is not only reflected but produced. Over the last few years, an increasing number of photographs depicting tourists and refugees on the same stretch of sand have appeared in the press. Sometimes, in these, tourists bask in the sun as corpses wash ashore, but sometimes holidaymakers try to rescue shipwrecked immigrants or volunteers from different countries assist local communities in alleviating the plight of refugees. On Europe's Mediterranean beaches as well as on the beaches of the British isles, and on beaches across the globe, littoral space is becoming simultaneously part of a reinforced border system and a site were the idea of fixed, neatly divided territories is being challenged: a heterotopia of difference indeed.

Notes

1 In the older photographs (by Tony Ray-Jones and David Hurn) the beach visitors are almost exclusively white, whereas in the more recent ones (by Parr and Simon Roberts) British beaches have become multicultural.
2 For a history of the beach as a (Western) vacation site, see Urbain (2003); Lenček and Bosker (1998).
3 While the beach is receiving progressively more attention in cultural and literary studies, it is hardly ever addressed as a possible heterotopia. Abdulrazak Gurnah's promisingly titled "Writing the Littoral" in a volume on globalisation and heterotopia discusses the coastal region of East Africa as "the western extent of the Indian Ocean world" (2015, 98) but does not really engage with its specific properties as an Oceanic rim, nor does Gurnah mention the shore or beach specifically.
4 As historical research has shown, the rise of seaside resorts as mass destinations was enabled by the railway connecting the coast to industrial centres; as the most advanced industrial country, Great Britain spearheaded this development in the nineteenth century. See Lenček and Bosker (1998, 107–9); Shields (1991, 85); Hassan (2003, 36).
5 The exclusion of particular groups, especially racial segregation, was a long-standing practice in various countries, for example the south of the United States or Apartheid South Africa. On resort antisemitism, the banishing of Jews from Germany's Baltic beaches, see Bajohr (2003). For literary engagements with segregation on the beach see, for instance, Samuelson (2015).

Works cited

Agamben, Giorgio. 2005. *State of Exception*. Translated by Kevin Attell. Chicago: University of Chicago Press.

———. 1998. *Homo Sacer: Sovereign Power and Bare Life*. Translated by Daniel Heller-Roazen. Stanford: Stanford University Press.

Arendt, Hannah. 1998 [1958]. *The Human Condition*. 2nd ed. Chicago: University of Chicago Press.

Augé, Marc. 1995. *Non-Places: Introduction to an Anthropology of Supermodernity*. Translated by John Howe. London and New York: Verso.

Bajohr, Frank. 2003. *"Unser Hotel ist judenfrei": Bäder-Antisemitismus im 19. und 20. Jahrhundert*. Frankfurt am Main: Fischer.

Cenzatti, Marco. 2008. "Heterotopias of Difference." In *Heterotopia and the City: Public Space in a Postcivil Society,* edited by Lieven De Cauter and Michiel Dehaene, 76–85. Abingdon: Routledge.

Certeau, Michel de. 1988. *The Practice of Everyday Life*. Berkeley and Los Angeles: University of California Press.

Cleave, Chris. 2016. *The Other Hand*. Reprint. London: Sceptre.

Colebrook, Claire, and Jason Maxwell. 2016. *Agamben*. Cambridge: Polity Press.

De Cauter, Lieven, and Michiel Dehaene. 2008. "The Space of Play: Towards a General Theory of Heterotopia." In *Heterotopia and the City: Public Space in a Postcivil Society*, edited by Lieven De Cauter and Michiel Dehaene, 87–102. Abingdon: Routledge.

Dehaene, Michiel, and Lieven De Cauter. 2008. "Introduction: Heterotopia in a Postcivil Society." In *Heterotopia and the City: Public Space in a Postcivil Society*, edited by Lieven De Cauter and Michiel Dehaene, 3–9. Abingdon: Routledge.

Denise-Benn, Nicole. 2017. *Here Comes the Sun*. London: Oneworld.

Foucault, Michel. 1986. "Of Other Spaces." *Diacritics* 16 (1): 22–27.

Garland, Alex. 1996. *The Beach*. London: Viking.

Gurnah, Abdulrazak. 2015. "Writing the Littoral." In *The Globalization of Space: Foucault and Heterotopia*, edited by Mariangela Palladino and John Miller, 95–110. London: Pickering and Chatto.

Hassan, John. 2003. *The Seaside, Health and the Environment in England and Wales Since 1800*. Aldershot: Ashgate.

Lenček, Lena, and Gideon Bosker. 1998. *The Beach: The History of Paradise on Earth*. New York: Viking.

Owens, Patricia. 2009. "Reclaiming 'Bare Life'? Against Agamben on Refugees." *International Relations* 23 (4): 567–82.

Parr, Martin. 2018. "Southend-on-Sea, Essex, 2017." In *The Great British Seaside: Photography from the 1960s to the Present*, Exhibition catalogue, 77. London: National Maritime Museum Greenwich.

Parr, Susie. 2018. "Martin Parr: The Essex Seaside, 2017." In *The Great British Seaside: Photography from the 1960s to the Present*, Exhibition catalogue, 76. London: National Maritime Museum Greenwich.

Perfect, Michael. 2014. *Contemporary Fictions of Multiculturalism: Diversity and the Millennial London Novel*. Basingstoke: Palgrave Macmillan.

Samuelson, Meg. 2015. "Literary Inscriptions on the South African Beach: Ambiguous Settings, Ambivalent Textualities." In *The Beach in Anglophone Literatures and Cultures*, edited by Ursula Kluwick and Virginia Richter, 121–38. Farnham: Ashgate.

Shields, Rob. 1991. *Places on the Margin: Alternative Geographies of Modernity*. London and New York: Routledge.

Soja, Edward W. 1991. *Thirdspace: Journeys to Los Angeles and Other Real-and-Imagined Places*. Oxford: Blackwell.

———. 1989. *Postmodern Geographies: The Reassertion of Space in Critical Social Theory*. London and New York: Verso.

Southend-on-Sea Borough Council. n.d. "Events and Leisure." Accessed January 31, 2019. www.southend.gov.uk/info/200305/events_and_leisure.

Southend Rising. n.d. "About Southend." Accessed March 25, 2019. www.southendrising.com/about-southend.html.

Urbain, Jean-Didier. 2003. *At the Beach*. Translated by Catherine Porter. Minneapolis and London: University of Minnesota Press.

9 Airbnb as an ephemeral space

Towards an analysis of a digital heterotopia

Elham Bahmanteymouri and Farzaneh Haghighi

Introduction

The main objective of this chapter is to argue that Airbnb is an ephemeral phenomenon which serves as an example of how heterotopian spaces can be rethought in the light of twenty-first-century global capitalism and the sharing economy. To achieve this objective, we present an investigation of the sharing economy with a focus on Airbnb as a popular digital platform in this new mode of economy. The sharing economy cannot be considered as an alternative to capitalism but as an innovative space for its renewal. This chapter employs the Foucauldian concept of heterotopia to explain how political, economic, and social relations have created Airbnb as a space that challenges prevailing norms and established relations.

By applying a Marxian-Lefebvrian interpretation, we challenge the way in which heterotopia is often understood, that is, as *a space for resistance to power*. Because heterotopias are temporary and they cannot last long, they may emerge as alternative spaces but cannot sustain their difference. Drawing upon Schumpeter's theory of creative destruction, this chapter argues that due to its creative nature, capitalism generates innovative spaces to challenge previously established norms. Such innovative spaces appear to be *anti-capitalist spaces of resistance*; however, they are already adopted by capitalism to overcome its crisis and they remain profit-making machines. Consequently, we stress that the heterotopic characteristic of Airbnb will disappear with its compliance with the norms and should be conceived of as an ephemeral space.

A Lefebvrian-Foucauldian interpretation of the sharing economy and Airbnb provides a critical understanding of this new mode of economy as a heterotopic and ephemeral space shaped by the current economic and social situation as well as by technological advancement to assist capitalism in solving economic crises. Although this new phenomenon represents a heterotopic space which is disrupting the rental housing market, government institutions, and the culture of ownership, it may disappear or change into a new space as soon as economic and social situations change. Central to our criticism is a specific Marxian political-economic stance that regards the rhetoric of *sharing* to be a mask hiding the aggressive aspects of Airbnb. This chapter sits completely in opposition to any perspective that blindly and solely praises Airbnb for its anti-hegemonic characteristics.

Airbnb, the sharing economy, and neoliberalism

This section explores how neoliberalism shaped the rationales for the sharing economy and allowed Airbnb to emerge. Capitalism as a creative mode of economy produces new areas of economic productivity to overcome its limitations and contradictions in order to survive. This creative mode of economy offers temporary economic solutions, ephemeral spaces, and social relations to deal with economic crises. As Marx explained in 1848, capitalism

> cannot exist without constantly revolutionising the instruments of production, and thereby the relations of production, and with them the whole relations of society. . . . All fixed, fast-frozen relations, with their train of ancient and venerable prejudices and opinions, are swept away, all newformed ones become antiquated before they can ossify. All that is solid melts into air, all that is holy is profaned.
>
> (Marx and Engels 2008, 25)

Following this Marxian notion, Schumpeter (2010) argued in 1942 that capitalism operates based on business cycles and periods of boom and bust; every round of economic growth is doomed to fall in an economic downturn, resulting in poverty and unemployment. The system has to recover itself through disruptive innovation and incessant creation and change. Capitalism is

> the . . . process of industrial mutation that incessantly revolutionizes the economic structure from within, incessantly destroying the old one, incessantly creating a new one. This process of Creative Destruction is the essential fact about capitalism. It is what capitalism consists in and what every capitalist concern has got to live in.
>
> (Schumpeter 2010, 83)

Based on Schumpeter's discussion, a strong relationship can be seen between the boom and bust of market operation and technological innovation (Kamien and Schwartz 1982). The incessant "creation of novelty and the destruction of old products and processes" (Andersen 2009, 162) occur when technological innovations deconstruct long-standing arrangements and free resources to be deployed elsewhere. At this point, capitalism needs "the new commodity, the new technology, the new source of supply, and the new type of organization" (Schumpeter 2010, 83). Complementary to technology, entrepreneurship is also an important element of Schumpeter's theory: it is the entrepreneur who facilitates the emergence of the new economy by managing both labour and natural resources.

The concept of creative destruction helps in understanding how the sharing economy operates. The sharing economy can be defined as "an economic activity in which web platforms facilitate peer-to-peer exchanges of diverse types of goods and services" (Aloni 2016). Schor (2016) argues that the sharing economy

> covers a sprawling range of digital platforms and offline activities, from financially successful companies like Airbnb, a peer-to-peer lodging service,

to smaller initiatives such as repair collectives and tool libraries. Many organizations have been eager to position themselves under the "big tent" of the sharing economy because of the positive symbolic meaning of sharing, the magnetism of innovative digital technologies, and the rapidly growing volume of sharing activity.

(7–8)

Airbnb's digitally mediated peer-to-peer platform can be viewed in light of Schumpeter's focus on technological innovation and the development of a new form of entrepreneurship. Furthermore, anticipating our argument to come, we can already suggest that Airbnb can be viewed as a site for Schumpeter's creative destruction, which has heterotopian qualities.

Within the context of the sharing economy generally, we maintain that the emergence of Airbnb should be seen against the background of three specific elements coming together: the 2007–8 economic crisis, technological advancement, and socio-cultural relations. The 2007–8 economic crisis is the first and main factor in the emergence of the sharing economy (Tuttle 2014; Hamari, Sjöklint, and Ukkonen 2015; Cockayne 2016). As the hegemonic ideology since the early 1980s, neoliberalism has been operating through a combination of innovative financial mechanisms alongside housing policies (Harvey 2014). However, this combination caused big failures that ended in the 2008 global economic collapse (Kotz 2015). The housing and financial crises led the US into recession, which consequently squeezed the economic situation of the middle class, with widespread unemployment exacerbating the economic situation. The large number of foreclosures crippled the financial and banking systems; they also decreased individuals' credit resulting in less loan and financial activities, which made it difficult for people to buy houses and for homeowners to rent out their houses. Due to the high rate of unemployment and high debt, people were forced to work in part-time or temporary jobs. Over the next few years, many people in the US who were suffering from the financial and housing crisis and were desperate and unemployed advertised spare rooms or spaces in their homes online. In many American cities, freelancers, seasonal contractors, and the unemployed started subletting from their landlords and renting out their bedrooms while sleeping on the couch in order to help pay the rent. The sharing economy with all its technological opportunities offered a means to replace destroyed jobs, relationships, and organisations by creating new ones to ensure recovery from the recession (Sundararajan 2016).

Airbnb therefore emerged as an alternative solution to save neoliberalism. In 2016, *The Guardian* reported that Airbnb claims to be "a solution to the problem of growing middle-class inequality that Donald Trump campaigned on, as it attempts to persuade local governments around the world of what it has to offer" (Hunt 2016). Chris Lehane, head of Airbnb global policy, specifically highlights that at the time, "opportunities presented by the sharing economy were helping to mitigate the blows dealt by the financial crisis in 2008" (Hunt 2016). Using idle items and bringing them back into the economic cycle helped in the recovery from economic downturn by providing opportunities for economic growth.

Second, technological advancement and the provision of internet access to an increasing number of people are considered another development feeding into the rise of the sharing economy (Taeihagh 2017; Codagnone, Karatzogianni, and Matthews 2018). Based on Schumpeter's concept of creative destruction (Schumpeter 2009, 2010), innovation and new technology can be seen as important factors for economic transformation and for a new mode of economic productivity in times of economic downturn.

Third, a combination of economic crisis and technological advancement resulted in a set of social relations which have attracted many people to this new mode of economy in order to take advantage of it (Hamari, Sjöklint, and Ukkonen 2015). Hence, being a consumer within the sharing economy, in particular Airbnb, needs active participation in a specific set of social relations that involves the use of technology in daily life. This may explain why most of the demand in the sharing economy, particularly Airbnb, comes from technologically literate millennials rather than baby-boomers (Williams 2018). Following Žižek (2009), O'Regan and Choe (2017) draw on the notion of *cultural capitalism* and argue that the sharing economy offers "a new stage of commodification that does not change the basic rules of capitalism. . . . We no longer buy products to own, but seek life experiences to render life meaningful" (O'Regan and Choe 2017, 1). Indeed, the culture of the sharing economy is shaped by market relations, which are enhanced through using smart devices that sustain the operation of this mode of economy.

These technologically mediated social relations that emerge through Airbnb also link the sharing economy with tourism. Besides offering "residents the opportunity to earn extra income by renting out their homes," Airbnb promises "tourists authentic and 'off-the-beaten-track' experiences of staying with locals" (Nieuwland and Van Melik 2018, 1) as a new trend in tourism. The Airbnb slogans invite tourists to temporarily settle in local people's homes and experience the local culture. The slogans "refer to hypothetical real and trustworthy travellers entering a contact zone together with equally real and trustworthy 'local' subjects" (Roelofsen and Minca 2018, 172).

Sharing economy criticism

Although the sharing economy and Airbnb are new phenomena, there is a large body of literature that has investigated their economic, environmental, social, and cultural dimensions. Obviously, Airbnb has provided a way out of the economic crisis: it has created new job opportunities, provided access to cheap and short-term accommodation for lower income groups, and consequently created money and contributed to economic growth. Some literature even contends that the sharing economy has moved beyond the capitalist economy towards socialism (Rifkin 2014; Richardson 2015). However, Airbnb has received much criticism. Among the fast-growing body of literature, studies that have taken a *qualitative* approach in order to problematise the logic of this online platform are relatively scarce. We categorise the criticism of the sharing economy into two main issues.

First, a frequent point of criticism is the sharing economy's suggested new approach beyond capitalism. Cockayne (2016) suggests that the sharing economy, which is a capitalist strategy for circulating fixed-capital more effectively, is in fact paradoxical because it promises capitalism and its alternative simultaneously. He casts doubt on the romanticised characteristics of sharing by pointing to the exploitative aspects of such platforms. For example, the labour of the Uber driver or an Airbnb host is disposable and they are required to "respond immediately to the demands of smartphone users" (Cockayne 2016, 80). The distinction between service providers and users reflects the emergence of a social class difference in third-party digital platforms (Irani 2015). In particular, the new technological advancement and digital contexts of capitalism often justify on-demand, low-paid contracts instead of permanent employment. This means that conditions of minimum payment and workers' compensation for overtime can be avoided, while also undermining "workers' possibilities for collective action and progressive changes to their circumstances" (Cockayne 2016, 80).

According to Roelofsen and Minca (2018), Airbnb has established an idealised, nostalgic, and allegedly lost notion of community, claiming to offer tourists authentic and local experiences and the ability to belong anywhere. This is evident in Airbnb's billboard or newspaper advertisements – "we imagine a world where you can belong anywhere" (Passiak 2017). However, the authors problematise this rhetoric by pointing out how the intimate domain of private life has been opened up to commercial interest and quantified by ranking. An Airbnb host is expected to be an expert on the local neighbourhood and to facilitate an encounter whereby the guest feels at home, while opening up their private space to the gaze of a stranger. This encounter has a biopolitical dimension, as the bodily performance of the host is constantly under a ranking mechanism leading to an algorithm that generates a digital community which excludes undesirable behaviours, bodies, and places (Roelofsen and Minca 2018). In this manner, the ranking system defined by the rationality of the market creates competitiveness; however, Airbnb obscures the brutality of the market competition by creating a fantasmatic image of a community.

Second, critics argue that the revenue from the sharing economy should be included in the local and governmental/official economy. Airbnb has been seriously criticised for tax evasion. O'Regan and Choe explain that "Airbnb also throws into question whether a future self-interested consumer (many of whom do not currently pay stipulated room-tourist taxes through Airbnb) will pay into any tourist accommodation tax system, which often supports public services such as environment preservation, transport, and even education" (2017, 169).

In addition, data-driven urban planning studies have emerged that analyse the impact of Airbnb on the supply and affordability of residential rentals and on local communities in specific cities. Simply stated, it can be argued that listing properties on Airbnb leads to a reduction of houses in the residential rental market, and consequently local rental rates increase. Local residents, unable to afford rent or to buy a house, particularly in areas with tourist attractions, move out, and these neighbourhoods turn into blocks of hotels. In many cities around the world,

Airbnb has been criticised for not paying tax to local government and for being one of the variables contributing to housing unaffordability (Lee and Rev 2016). In 2018, the Auckland Council (Tuatagaloa and Osborne 2018) identified Airbnb as one of the key variables of unaffordability in the rental housing market. The new challenge for the government is therefore not only taxes, but also the gentrification and displacement of neighbourhoods that are turning into accommodation for tourists rather than supporting the local community.

Described as disruptive by the tourism and hospitality literature, Airbnb has also been identified as a competitor that is threatening the future of the hotel and tourism industry (Guttentag 2013; Oskam and Boswijk 2016; Zervas, Proserpio, and Byers 2017). Furthermore, Airbnb has also been criticised for biased listings. For example, in a study of two Australian cities, Sydney and Melbourne, Alizadeh, Farid, and Sarkar (2018) found that the locational patterns of Airbnb listings are influenced by race, income, and education. These factors impact the economic activities on the platform, meaning that the lowest end of the socio-economic spectrum of the population has the least financial gains. Airbnb is thus unlikely to generate income for low earners (Gurran and Phibbs 2017).

Ultimately, these critical studies call for regulatory intervention, which is already present in some European and North American cities where Airbnb is regulated, such as Amsterdam, Anaheim, Barcelona, Berlin, Denver, London, New Orleans, New York, Paris, San Francisco, and Santa Monica (Nieuwland and Van Melik 2018).

Airbnb as an ephemeral space

We argue that the sharing economy, in particular Airbnb, has an ephemeral essence, which is also where the connection to heterotopia lies. The meteoric rise of Airbnb was a consequence of the economic downturn and the squeezing of the middle class; however, it has been argued that economic recovery will end this temporary phenomenon (Oskam and Boswijk 2016). It will metamorphose and transform into something else as soon as the developments that made its emergence possible change. In architectural and urban discourse, ephemerality is mostly used to refer to the temporal dimension of a fleeting situated event. The notion of the ephemeral has often been employed as a descriptive term rather than an analytical concept for spatial analysis (Cowan 2002; Baek 2006; Hill 2006; Bonnemaison and Macy 2008); however, for urban design scholar Madanipour (2017a), this concept plays a more critical role. Discussing the temporal use of urban space, Madanipour highlights that the instrumental treatment of temporality created and accelerated by digital technologies has led to the sense of restlessness and ephemerality experienced by city dwellers. Following the promise of modernity to create new spaces (through densification and relocation) and to reduce time (through speed and multiplication), the world is now rendered as a small village that is accessible at any time and without any limits to the human experience (Madanipour 2017a). How modernity has radically changed the way time and space are experienced is also explained by Harvey (1990) through the notion of *time-space compression*. For Harvey, modernist innovations in transportation and communication technology provided the material foundation for accelerating the turnover time in production, exchange, and consumption whilst reducing the space between diverse communities across the globe. In this rapidly changing society, spatial barriers and time horizons collapsed and individuals were "forced to cope with disposability, novelty and the prospects for instant obsolescence" (1990, 286). They were also forced to live in close proximity with each other while competing to make their locale more attractive to highly mobile capital (Harvey 1990, 295).

Cities within this contemporary condition grew and intensified at unprecedented rates. However, while transformed into "places of temporal densification and magnification," the limits of cities were revealed and globalisation, as a process of relocation, emerged to overcome these temporal and spatial constraints with a promise of reducing costs and increasing profits (Madanipour 2017a, 49–50). The flexibility of the labour market (part-time, casual, and non-contractual jobs) also contributed to this fast expansion because "when the conditions of work can change at higher speed, it may provide the possibility of expanding or contracting time as and when needed, according to the conditions of the market" (Madanipour 2017a, 50).

For Madanipour (2017a, 72), the resulting temporary urbanism has generated a sense of aspatiality, immateriality, and ephemerality. Referring to the growing number of transient events (such as pop-up shops, street festivals, mobile buildings, and temporary gardens) aimed at animating de-industrialised vacant spaces, he argues (2017b) that these are dramatic conceptual changes as they introduce ephemerality as opposed to durability. These interim spaces used for transient events are temporary and the intensification of ephemerality is extended to the city's physical infrastructure. Consequently, the urban space that was historically understood as a stable setting is now the subject of continuous and everchanging utilitarian considerations. The urban landscape itself has become an event that favours ephemerality.

The concept of an event, particularly as used in architecture, expands our understanding of the ephemerality of Airbnb. An event offers a different experience of space, and Airbnb operates upon this particular experience. In the 1980s, Bernard Tschumi identified the shift in architectural discourse by highlighting the significance of events over the supremacy of space or form, stating that

> architecture should be defined not only as space but also as the thing that happens in the space, that is, as space and event. As soon as you start talking about the thing that happens in the space, you enter a social and therefore political dimension.
>
> (Hartoonian 2002, 79)

The event itself is short-lived, ephemeral, and unpredictable, and it disrupts or reshuffles the existing structures (Haghighi 2018, 161). Airbnb operates as an event because it disrupts the norm of the hospitality industry and provides short-lived experiences in urban space. As an event, Airbnb is ephemeral because of its inherent relation to capitalist rationality and the "volatility and ephemerality of fashions, products, production techniques, labour processes, ideas and ideologies, values and established practices" (Harvey 1990, 285).

While historically, houses have been associated with stability, privacy, and belonging, an Airbnb house is opened up as an ephemeral space for a stranger to pass through. When a stranger enters the house, it becomes an experience in which the distinction between a series of binary oppositions, such as private/public, resident/tourist, and community/business, becomes blurred. However, such boundaries do not disappear; they remain operative and create an illusion of being included in the family or the community. This reminds us of Foucault's reference to heterotopic sites that hide curious exclusions – for example, Brazilian farmhouses that accommodate guests inside the house with a clear separation from the family's quarters and one remains only a guest in transit (Foucault 1986, 26).

By using existing houses and operating in the digital space of the internet, Airbnb renders the binary opposition of virtual/physical obsolete; as Grosz (2001, 41) suggests, cyberspace is always located within a real space. Through the use of digital technologies, Airbnb connects disjointed elements, places, and times and creates a hybrid environment. Airbnb creates an ephemeral space within houses, and as a digital platform it enables digital space to impact the physical realm through a number of transformations that range from the renovation of individual houses to the displacement of permanent accommodation in some neighbourhoods. In sum, it can be argued that ephemerality refers to a temporary event, which reconfigures the physical settings in favour of a short-lived and commercialised experience.

A Lefebvrian interpretation of the heterotopic space of Airbnb

We draw on a Lefebvrian interpretation of heterotopia to understand and analyse how Airbnb appeared as a heterotopic space with ephemeral characteristics that has challenged established institutional and cultural norms. Although heterotopic spaces emerge as a necessity for the sake of the creative destruction of capitalism, they do not offer any space to resist the dominant capitalist relations. They are trapped; they will be destroyed and will be replaced with another form, solely to guarantee the incessant process of accumulation of capital. Indeed, Lefebvrian heterotopia is the materialisation of Schumpeter's theory of the renewal of capitalism in an urban space. In this section we explain how heterotopias are ephemeral and will transform to a normal space deprived of any potential for resistance.

Lefebvre (1996, 2003) outlines a history of urbanisation and presents a novel definition of urban space, not only in terms of its connection with the critical and political role of the city, but also as a dialectical, contradictory, and evolutionary process of tensions, conflicts, and juxtapositions, and as a place reflecting the reality of the politics of everyday life with a transforming potential to claim the political rights of its habitants. For Lefebvre, space is characterised by the use attributed to that particular space; for example, public space is the process and the act of making things public. Importantly, he criticises the roles and responsibilities of architects, planners, and urbanists who work in the ideological *blind field* of urbanism and spatial studies (1991, 2003). He argues that planners and architects work in an abstract ideological space which results in a reductionist reading of space that is blind to many unquantifiable constituents of space as a social product. According to Lefebvre (2003), spaces are marked by different superimposed functions and can be categorised into three types of space (topoi): isotopies, heterotopies, and utopias. These spaces "coexist, interact and are produced in relation to one another" (Cenzatti 2008, 80). Lefebvre contends that these three concepts and their dialectical relations provide a better understanding of the complexity of urban space under the capitalist mode of production. For our purposes of exploring the ephemerality of Airbnb, the relation between heterotopies and isotopies is particularly relevant.

Isotopies, places of identity or identical spaces, are spaces of order and are created by the state of rationalism. "An isotopy is a place (*topos*) and everything that surrounds it (neighbourhood, immediate environment), that is, everything that makes a place the *same place*" (Lefebvre 2003, 37–38). Isotopies are homogenised spaces: everything is calculable, predictable, and quantifiable. Everything is part of an order, such as residential and/or office blocks, arrondissements, and the bureaucratic limits of electoral districts, as well as long straight lines, broad avenues, and empty perspectives without regard for either the rights or interests of the lower class or cost (Lefebvre 2003, 128–31). Isotopies are comparable spaces that can be discussed and read through maps and images.

Heterotopies are "the place of the other and the other place simultaneously excluded and interwoven" (Lefebvre 2003, 128). For Lefebvre, heterotopies are interwoven fields of time-space. In contrast to Foucault, Lefebvre "envisaged heterotopias in a more critical register, rooting them in a sense of political and historical deviance from social norms" (Smith 2003, xii). Illustrated in spaces such as caravansaries, fairgrounds, and suburbs, heterotopias operate independently from the places of official commercial exchange.

Although isotopy and heterotopy appear differently, in that the former is related to established norms while the latter indicates ephemeral events that challenge established social, institutional, as well as cultural norms, they are intertwined. According to Lefebvre (2003, 191), "The description of isotopies and heterotopies goes hand in hand with the analysis of the acts and situations of the individual and collective subjects and their relation to objects populating the urban space." Heterotopies are ephemeral and transform into isotopic spaces as soon as they can be read, comprehended, calculated, and mapped. "The isotopy-heterotopy difference can only be understood dynamically. In urban space, something is always happening. Relations change. Differences and contrasts can result in conflict, or are attenuated, erode, or corrode" (Lefebvre 2003, 129). Lefebvre explains that urban space has always been heterotopic. In European cities of the sixteenth and seventeenth centuries, suburban areas could be considered heterotopic spaces in that they were equipped with poor infrastructure and were occupied by "traders, mercenaries, workers, and the poor, or people who often [were] suspect and sacrificed in time of war" (Lefebvre 2003, 129). Later, when the urban services and facilities expanded to the peripheral heterotopies, these areas turned into isotopies. Therefore, there is always a dialectical relationship between isotopies and heterotopies.

Following a Lefebvrian interpretation of space, the dialectical relationship between isotopy and heterotopy explains how the sharing economy, in particular Airbnb, emerged as a heterotopic space that challenged the social and institutional norms. Airbnb offered a new experience of travelling and communication between participants in the tourist industry. Participants that were functionally engaged in this space became strangers to those who were not. In this sense, Airbnb was a heterotopic space in its initial stage, when it emerged to offer a new technological space of exchange beyond the traditional methods of communication. It established a new marketplace beyond traditional space and the time axis; everyone could participate in this market without tax and other normal hindrances of decrees and orders. Hence, Airbnb created a space for exchange and trade that was initially marked by the signs of heterotopy.

From an institutional point of view, hotels, homestays, and/or other traditional and local/national platforms and institutions have always supplied temporary accommodations for tourists. These platforms are clearly known and registered; therefore they pay tax on their share of capital gains from urban space. Similarly, participants from the demand side of the tourism market pay direct or indirect tax as their contribution to local governments. Airbnb offers a new space of interaction between demand and supply in this market, a space that operates as a non-officially registered space that evades paying tax and creates a noticeable benefit for people who are engaged in this space.

Based on the Lefebvrian interpretation of dialectical relations between spaces, Airbnb will become part of the isotopic space of the rational order as soon as it can be taxed, regulated, and taken for granted as a popular platform for temporary accommodation. The life span of Airbnb, it can be argued, comprises two stages. First, it emerged as an innovative and disruptive space to challenge established but less profitable norms and relations, such as the tourism industry. Second, it will eventually comply with the tax system of each country or city and therefore transform into an isotopic space. As soon as companies within the sharing economy, such as Airbnb, follow regulations and state norms, they can no longer produce as much profit as achieved in their initial stages; consequently, the productive contribution to global capital growth decreases (Oskam and Boswijk 2016), even though the contribution to the local economy increases. Based on the notion of creative destruction, initiatives are doomed to this business cycle. Capitalism, therefore, needs new innovative market spaces to overcome another round of economic downturn.

In summary, the Lefebvrian interpretation of heterotopia clarifies that sharing-economy platforms such as Airbnb, which initially emerged as a heterotopic space of exchange, are now transforming into isotopic spaces of rational order. The process of heterotopic spaces transforming into isotopic ones is part of the probable disappearance of Airbnb, which itself is a phase in a bigger cycle of capitalist metamorphosis.

A Foucauldian interpretation of the heterotopic space of Airbnb

In line with Lefebvre, Foucault (1986, 24) asserts that alternative real spaces can emerge which are "in relation with all the other sites, but in such a way as to suspect, neutralize, or invert the set of relations that they happen to designate, mirror, or reflect." Whilst discussing the conditions necessary for the emergence of modern knowledge in the preface to *The Order of Things*, Foucault (2007) introduces heterotopia in contrast with utopia as a *disturbing* condition. The perfect organisation of things and words in utopia cannot occur in heterotopia because the structural order through which language and things are linked together no longer exists. Such a situation is disturbing because multiple orders can be expressed and therefore established concepts are reconceived as historically situated constructions. For example, the concept of Man is a recent invention only two centuries old, and "a new wrinkle in our knowledge" that will disappear again "as soon as that knowledge has discovered a new form" (Foucault 2007, xxv). Through binary oppositions, Foucault describes heterotopic situations as not the ideal world, but rather the instability of the ground stirring under our feet.

Following Foucault, we argue that a heterotopic condition is ephemeral, allowing the multiplicity of possibilities to be expressed in contrast to the accepted norms and ideals. Thus, it is no surprise that Airbnb is based on a different model that aims to unlock economic opportunities (AirbnbCitizen 2019) and provide an alternative way to travel. As described explicitly in the "About Us" section on the website: "Airbnb *uniquely* leverages *technology* to *economically* empower millions of people around the world to unlock and monetize their spaces, passions and talents to become hospitality entrepreneurs" (Airbnb 2019, emphasis added). That is, Airbnb offers an accessible way out of the severe economic crisis.

By injecting difference into the tourism industry, Airbnb operates in a similar way to a heterotopia, both through its discursive and spatial structure. As noted by Wesselman (2013), physical spaces are intertwined within the complexity of the urban fabric and are unable to suspend existing spatial structures; however, it is within language that an alternative space can continue to operate. As an alternative space, Airbnb also calls for a rethinking of this Foucauldian concept because of its multiple functions at the level of language (rhetoric of community), physical space (idle spaces), and economics (rescuing the middle class). Within this system, guests and hosts are framed as citizens of *another* community, an individual's private space is defamilarised to create a *different* type of home, and hosts are promised an *alternative* source of income. Airbnb is therefore framed as an alternative, different, or other space at a discursive level. In language, a counter-space can oppose dominant norms, but when it comes to real physical space, Airbnb remains embedded within the existing structures. It is not absolutely anti- or counter- and it cannot suspend all the socio-economic relations. As explained earlier, even at the level of economics, Airbnb does not operate differently to the capitalist mode of production; rather, it is embedded in it and perpetuates its operation.

For Foucault, heterotopias are real spaces linked and intertwined with other spaces whilst contradicting them – they are counter-spaces. Characterised as a disruptor of the hotel industry – here disruptive does not mean resistance in a Marxian sense – Airbnb deviates from traditional tourism and the notion of belonging to a single place and as such it can be easily aligned with six principles that Foucault explicitly lists for heterotopias in "Of Other Spaces" (1986). The function of Airbnb since the financial crisis of 2007–8 has transformed: it began by offering affordable rental housing before developing into a billion-dollar company that offered luxurious accommodation, and more recently, experiences and restaurants. Several incompatible spaces have been brought together in this digital heterotopia: home/hotel, community/business, virtual/physical, and familiar/unfamiliar. Access to this hospitality service is mediated via ID verification, bank accounts, apps on smartphones, and as recent events have highlighted, race as well (Alizadeh, Farid, and Sarkar 2018). Those who enter and use the advertised spaces are offered heterotopic experiences. The website portrays an image of a meticulously decorated, organised, ideal home; however, as soon as the traveller enters that space, they are excluded from the host's personal territory. Airbnb therefore provides not only heterotopic experiences, but it is itself a heterotopia of deviance.

At this point, another aspect of the Foucauldian notion of heterotopia merits close attention, that is, the difference between heterotopia and other spatial configurations that Foucault studied in detail, such as prisons, asylums, hospitals, or the town. Such modern institutions have emerged to transform the excluded into productive citizens within the capitalist system. Heterotopia, however, is intertwined with the city and appears with an ability to problematise and disrupt a set of social, economic, political, and cultural relations. Heterotopias exist as long as they remain "absolutely different" (Foucault 1986, 24). Although Foucault does not discuss the life span of these different spaces, we argue that a heterotopic site is ephemeral and cannot last long. Due to social and cultural changes, divergence from the norm is in constant flux. For example, various excluded or repressed groups have been recognised and integrated into society; they are the norm now and part of the machinery of production (Cenzatti 2008). Heterotopias fluctuate between acceptance/contradiction and recognition/invisibility. For instance, how long can a cemetery remain a counter-site and resist becoming a touristic attraction generating profit for a city (Clements 2017)? How long can countercultures retain their difference when faced with the apparatus of urban governance that seeks to transform them into a brand for a creative city (Scott 2016)? Sooner or later, they will be appropriated into the norm and commercialised.

What matters to Foucault is that an alternative space can emerge, rather than discussing how long this difference can last. Lefebvre, however, identifies the ephemerality of heterotopias in addressing the life span of difference by highlighting the dialectical relation between isotopies, heterotopies, and utopias. For Lefebvre, heterotopias, or different spaces, are ephemeral as they constantly transform into isotopies. Hence, spaces that are located outside of the system are doomed to be captured, measured, regulated, and recirculated within the capitalist city.

Concluding remarks: an ephemeral spatial opportunity for the metamorphosis of capitalism

In this chapter, we pointed out that Airbnb has heterotopic characteristics and as such it stands as a constructive case study to demonstrate how Foucault's original formulation of the notion of heterotopia can be extended to our contemporary, globalised world. By explaining that Foucauldian and Lefebvrian interpretations of heterotopia share similar principles, we suggested that heterotopic spaces such as the sharing economy and particularly Airbnb should be considered as creative and disruptive spaces deployed for the renewal of capitalism. In doing so, we challenged the common interpretation of heterotopia as *a space for resistance to power*.

Airbnb emerged to save neoliberalism from the 2007–8 economic crisis and it developed through advancements in digital technology. It grew to attract many more people, promising its users they could belong anywhere – a rhetoric clearly in line with the globalised culture of the sharing economy. In fact, Airbnb is inseparable from the global city, global capitalism, global corporations, the neoliberal entrepreneurialised workforce, and the global web-based community. Although Airbnb is globally considered a popular type of temporary accommodation – with more than 3 million active listings available across 191 countries, including 65,000 cities (Airbnb Press Room 2019) – it may remain heterotopic as long as it escapes formal regulations. In a sense, Airbnb reflects the liberating aspect of heterotopia, as it diverts from defined and fixed norms, such as tax and planning regulations.

Being heterotopic means the space should be seen as ephemeral. We argued that the heterotopic nature of Airbnb has two ephemeral characteristics. First, it offers a temporary event, which reconfigures physical settings in favour of a short-lived and commercialised experience. Second, in line with Lefebvre, Airbnb is destined to transform into an isotopic space as soon as it follows regulations. To conclude, Airbnb is an ephemeral space as it is a tool of the creative destruction mechanism. Capitalism always needs both innovative and disruptive spaces to overcome its limitations. For this reason, Airbnb is a fad that may disappear in the next round of economic boom/bust and the emergence of a new technology will become another heterotopic space.

Works cited

Airbnb. "What Legal and Regulatory Issues Should I Consider Before Hosting on Airbnb?" Accessed July 1, 2019. www.airbnb.co.nz/help/article/376/what-legal-and-regulatory-issues-should-i-consider-before-hosting-on-airbnb.

AirbnbCitizen. "About Airbnb Advancing Home Sharing as a Solution." Accessed July 1, 2019. www.airbnbcitizen.com/.

Airbnb Press Room. Accessed July 1, 2019. https://press.airbnb.com/about-us/.

Alizadeh, Tooran, Reza Farid, and Somwrita Sarkar. 2018. "Towards Understanding the Socio-Economic Patterns of Sharing Economy in Australia: An Investigation of Airbnb Listings in Sydney and Melbourne Metropolitan Regions." *Urban Policy and Research* 36 (4): 445–63.

Aloni, Erez. 2016. "Pluralizing the Sharing Economy." *Washington Law Review* 91: 1397–459.

Andersen, Esben Sloth. 2009. *Schumpeter's Evolutionary Economics: A Theoretical, Historical and Statistical Analysis of the Engine of Capitalism*. London: Anthem Press.

Baek, Jin. 2006. "Mujo, or Ephemerality: The Discourse of the Ruins in Post-War Japanese Architecture." *Architectural Theory Review* 11 (2): 66–76.

Bonnemaison, Sarah, and Christine Macy, eds. 2008. *Festival Architecture*. Abingdon: Routledge.

Cenzatti, Marco. 2008. "Heterotopias of Difference." In *Heterotopia and the City: Public Space in a Postcivil Society*, edited by Michiel Dehaene and Lieven De Cauter, 75–84. Abingdon: Routledge.

Clements, Paul. 2017. "Highgate Cemetery Heterotopia: A Creative Counterpublic Space." *Space and Culture* 20 (4): 470–84.

Cockayne, Daniel. 2016. "Sharing and Neoliberal Discourse: The Economic Function of Sharing in the Digital on-Demand Economy." *Geoforum* 77: 73–82.

Codagnone, Cristiano, Athina Karatzogianni, and Jacob Matthews. 2018. *Platform Economics: Rhetoric and Reality in the Sharing Economy*. Bingley: Emerald Publishing.

Cowan, Gregory. 2002. "Nomadology in Architecture: Ephemerality, Movement and Collaboration." PhD thesis, The University of Adelaide, Australia.

Foucault, Michel. 2007. *The Order of Things: An Archaeology of the Human Sciences*. Translated by Paul Rainbow. 8th ed. London: Routledge.

———. 1986. "Of Other Spaces." *Diacritics* 16 (1): 22–27.

Grosz, Elizabeth. 2001. *Architecture from the Outside*. London: MIT Press.

Gurran, Nicole, and Peter Phibbs. 2017. "When Tourists Move in: How Should Urban Planners Respond to Airbnb?" *Journal of the American Planning Association* 83 (1): 80–92.

Guttentag, Daniel. 2013. "Airbnb: Disruptive Innovation and the Rise of an Informal Tourism Accommodation Sector." *Current Issues in Tourism* 18 (12): 1–26.

Haghighi, Farzaneh. 2018. *Is the Tehran Bazaar Dead? Foucault, Politics and Architecture*. Newcastle upon Tyne: Cambridge Scholars Publishing.

Hamari, Juho, Mimmi Sjöklint, and Antti Ukkonen. 2015. "The Sharing Economy: Why People Participate in Collaborative Consumption." *Association for Information Science and Technology* 67 (9): 2047–59.

Hartoonian, Gevork, and Bernard Tschumi. 2002. "An Interview with Bernard Tschumi." *Architectural Theory Review* 7 (1): 79–88.

Harvey, David. 2014. *Seventeen Contradictions and the End of Capitalism*. Oxford: Oxford University Press.

———. 1990. *The Condition of Postmodernity: An Enquiry into the Origins of Cultural Change*. Cambridge, MA: Blackwell.

Hill, Jonathan. 2006. *Immaterial Architecture*. London: Routledge.

Hunt, Elle. 2016. "Airbnb a Solution to Middle-Class Inequality, Company Says." *The Guardian*, Tuesday December 13. www.theguardian.com/technology/2016/dec/14/airbnb-a-solution-to-middle-class-inequality-company-says.

Irani, Lilly. 2015. "Difference and Dependence Among Digital Workers: The Case of Amazon Mechanical Turk." *South Atlantic Quarterly* 114 (1): 225–34.

Kamien, Morton, and Nancy Schwartz. 1982. *Market Structure and Innovation*. Cambridge: Cambridge University Press.

Kotz, David. 2015. *The Rise and Fall of Neoliberal Capitalism*. Cambridge, MA: Harvard University Press.

Lee, Dayne, and Poly Rev. 2016. "How Airbnb Short-Term Rentals Exacerbate Los Angeles's Affordable Housing Crisis: Analysis and Policy Recommendations." *Harvard Law and Policy Review* 10: 229–53.

Lefebvre, Henri. 2003. *The Urban Revolution*. Translated by Robert Bononno. Minneapolis: University of Minnesota Press.

———. 1996. *Writings on Cities*. Translated by Eloenore Kofman and Elizabeth Lebas. Oxford: Blackwell.

———. 1991. *The Production of Space*. Translated by Donald Nocholson-Smith. Oxford: Blackwell.

Madanipour, Ali. 2017a. *Cities in Time: Temporary Urbanism and the Future of the City*. London: Bloomsbury.

———. 2017b. "Ephemeral Landscape and Urban Shrinkage." *Landscape Research* 42 (7): 795–805.

Marx, Karl, and Friedrich Engels. 2008. *The Communist Manifesto*. Edited by David Harvey and Samule Moore. London: Pluto Press.

Nieuwland, Shirley, and Rianne Van Melik. 2018. "Regulating Airbnb: How Cities Deal with Perceived Negative Externalities of Short-Term Rentals." *Current Issues in Tourism*: 1–15.

O'Regan, Michael, and Anatolia Choe. 2017. "Airbnb and Cultural Capitalism: Enclosure and Control Within the Sharing Economy." *An International Journal of Tourism and Hospitality Research* 28 (2): 163–72.

Oskam, Jeroen, and Albert Boswijk. 2016. "Airbnb: The Future of Networked Hospitality Businesses." *Journal of Tourism Futures* 2 (1): 22–42.

Passiak, David. 2017. "Belong Anywhere – The Vision and Story Behind Airbnb's Global Community." *Medium*, January 31. https://medium.com/cocreatethefuture/belong-anywhere-the-vision-and-story-behind-airbnbsglobal-community-123d32218d6a.

Richardson, Lizzie. 2015. "Performing the Sharing Economy." *Geoforum* 67: 121–29.

Rifkin, J. 2014. *The Zero Marginal Cost Society: The Internet of Things, the Collaborative Commons, and the Eclipse of Capitalism*. New York: St. Martin's Press.

Roelofsen, Maartje, and Claudio Minca. 2018. "The Superhost: Biopolitics, Home and Community in the Airbnb Dream-World of Global Hospitality." *Geoforum* 91: 170–81.

Schor, Juliet. 2016. "Debating the Sharing Economy." *Journal of Self-Governance and Management Economics* 4 (3).

Schumpeter, Joseph. 2010. *Capitalism, Socialism, and Democracy*. London: Routledge.

———. 2009 [1947]. *Can Capitalism Survive?* London: HarperCollins.

Scott, Felicity. 2016. *Outlaw Territories: Environments of Insecurity/Architectures of Counterinsurgency*. New York: Zone Books.

Smith, Neil. 2003. "Foreword." In *The Urban Revolution*, vii–xiii. Minneapolis: University of Minnesota Press.

Sundararajan, Arun. 2016. *The Sharing Economy: The End of Employment and the Rise of Crowd-Based Capitalism*. Cambridge, MA: The MIT Press.

Taeihagh, Araz. 2017. "Crowdsourcing, Sharing Economies and Development." *Journal of Developing Societies* 33 (2): 191–222.

Tuatagaloa, Penelope, and Brian Osborne. 2018. "Airbnb and Housing in Auckland." In *Auckland Council Technical Report*, 40. Auckland, New Zealand: Auckland Council.

Tuttle, Brad. 2014. "Can We Stop Pretending the Sharing Economy Is All About Sharing?" http://money.com/money/2933937/sharing-economy-airbnb-uber-monkeyparking/.

Wesselman, Daan. 2013. "The High Line, 'The Balloon,' and Heterotopia." *Space and Culture* 16 (1): 16–27.

Williams, Robert. 2018. "Forrester: Millennials Boost Growth of Sharing Economy." www.mobilemarketer.com.

Zervas, Georgios, Davide Proserpio, and John W. Byers. 2017. "The Rise of the Sharing Economy: Estimating the Impact of Airbnb on the Hotel Industry." *Journal of Marketing Research* 54 (5): 687–705.

Žižek, Slavoj. 2009. *First as Tragedy, Then as Farce*. London: Verso.

10 New communication technologies and the transformations of space

Lessons from Michel Serres' *Thumbelina*

Peter Johnson

Introduction

The idea for this chapter was sparked by Michel Serres' popular book *Thumbelina* (2015a) in which the philosopher mischievously and provocatively celebrates the new capabilities of young people brought up in the digital age. Although simply written as a modern fable, the book encompasses a range of methods and subjects developed in Serres' wider work. Three themes struck me as particularly relevant to the notion of heterotopia: how emerging digital communication technologies (DCTs) are producing a radically new global space; how young people in particular are being liberated by the opportunities of mobile technologies; and finally the wider potential inventiveness and alternative modes of thought opened up through the digital environment. The chapter explores how these transformations and potentialities, with accompanying dangers and threats, have an impact on, and engage with, the notion of heterotopia. If as Serres asserts we are in the midst of a shift in thought, imagination and experience that is at least as profound as the changes brought about by the invention of writing or the printing press, where does this leave heterotopia?

I begin the chapter by critically examining claims that some conceptions of a virtual world or the internet itself might be identified as heterotopia. Second, I turn to arguments that claim certain online sites such as Facebook or Wikipedia have heterotopian characteristics. Finding such claims both problematic and suggestive, I move to a wider perspective, investigating different ways in which heterotopia might be seen in the context of the interface between humans and technology or our overall interaction with our digital environment. As the chapter progresses, I increasingly take up thoughts from Michel Serres to argue for a repositioning of heterotopia in terms of identifying an emergent different space. In many ways my argument returns to the opening of Foucault's talk on heterotopia and stays there, engaging with his brief prescient remarks about our present era of "networks." A number of questions arise. How is our era different? What new opportunities emerge? How should we approach those opportunities inventively? In addressing these questions, I turn finally to Foucault's textual version of heterotopia as a stimulus for discovering creative ways to intervene in our new global epoch. I focus here specifically on addressing the emergency of our climate crisis and Serres' creation of Thumbelina, a teenager with her future literally in her hands, thumbing through her mobile phone.

An engagement with Serres' innovative and challenging work has been taken for a variety of reasons. The philosopher has always thought globally. In particular he focuses on the most recent stage of hominisation, which he calls "hominescence," the dawn of an era in which new technologies are transforming our relation to the world, ourselves, others, our bodies, time and space (Serres 2019, 9–11). This chapter will concentrate on the last transformation, the changes to our global environment mediated by the prevalence of advanced, mobile communication technologies and how we inhabit the world differently (193). Serres is also distinctive in imagining a hopeful future for this technology. Many academic publications focus on the clear dangers in terms of governments, political groups and global corporations controlling, monitoring and exploiting data and people through digital technologies, as summarised in a debate between Rouvroy and Stiegler (2016). In a discussion with his students, Serres (2017a) acknowledges these threats, particularly the damage to the democratic process through the manipulation of misinformation. Yet he provides a passionate alternative vision based on listening to the voices and observing the behaviour of young people, especially young women whom he sees as overturning entrenched divisions in terms of gender as well as starting to dismantle hierarchical structures in education, politics and culture (Serres 2016, 198).

I use certain strands of Serres' thought and methods as signposts and do not intend to cover the range of complex, innovative ideas and methods that cross his immense work. However, one of his underlying methods of study is worth introducing initially. Serres frequently plays with the idea of dismantling one of the core laws of logic, mathematics and thought, the "excluded third" which asserts that either a proposition is true, or its negative is true (2017b, 56–57). For Serres this type of reasoning leads into a series of restrictive binaries, for example order/disorder, subject/object, local/global, humanities/science, archaic/contemporary and virtual/real. For Serres, there is always a middle thought or perception, a mixture that includes being an integral part of the "noise," processes and fluctuations of the world (Serres 2017b, xliv). Importantly for this chapter, Serres identifies the earth itself as an excluded third. Through advanced technologies we are able to conceive for the first time the earth as a whole, as a partner (1995, 108–9). He imagines a global intuition that will allow the earth to be represented and "have a right to speak" (2014, 31–32). Finally, Serres explores the vibrant mixture of an inclusive world through methods that are similarly inclusive. As explored later, Serres works through "crumbled" thoughts and perspectives, opening up unanticipated and sometimes bewildering connections and syntheses.

The internet and online sites as heterotopia

This section will explore interrelated arguments that make a link between some notion of a virtual or digital space and heterotopia; first as an overriding equivalence and second as a fundamental quality of specific online sites. The first perspective is based on a broad correspondence with Foucault's descriptions of heterotopias, as claimed within the opening sentence of a recent article, "in the heterotopia that is the internet" (Leu 2018, 115). Where has this assumption come from? I think a main source stems from popular journal articles from the 1990s and the prevalent discourse of the time on the nature of "cyberspace." Those who equated cyberspace and heterotopia in this way often made a connection with Foucault's notion of a "placeless place" which occurs twice in his lecture in relation to the effect of the mirror as we face it (you are there and not there) and the journey of a ship criss-crossing the open sea (see Wark 1993; Young 1998). The argument suggests that in the first instance we are faced with something same and other, virtual and real, or tangible and abstract. The second compares the description of ships crossing between ports to the complex relations between terminals, or nodes in a network. Piñuelas applies both these models in his notion of "cyber-heterotopia" (2008).

The ship and its association with navigation does seem to have some resemblance to formulations of the internet or some virtual domain, particularly if we consider the etymology of "cybernetics" with its Greek roots implying steersman, pilot, or rudder as well as more recent metaphors of "navigating" or "exploring" the web or internet. In particular, the reference to the ship provides a handy bridge from the enclosed features of nearly all of Foucault's examples of heterotopia to an assumed expanse beyond the limitations of any geographic location. However, the major difficulty with this position concerns its attachment to notions of a distinctive, unbounded space or domain, "out there" as it were, a position thoroughly contested by geographers and researchers in new media. As Van den Boomen (2019) argues extensively, using all-embracing terms in this way retains if not reinforces misleading connotations concerning the diverse ways in which we interact with or inhabit digital technologies. The relation made to mirrors, on the other hand, may seem to resolve the idea of a separate domain but, as I go on to explore, actually preserves it through retaining a binary notion of the virtual and real, leading too easily into conceptions of the "other." The internet may have an important role in reimaging heterotopia, but it needs to be examined closely in the context of the possibilities opened up in an emerging and immersing digital environment.

Another line of enquiry turns away from conceiving heterotopia as our interactions with a virtual world and investigates spatio-temporal shifts of specific online sites that attract billions of users globally. Such sites are explored as offering a different mode of practice or experience that corresponds to or even intensifies Foucault's portrayal of heterotopia. For example, Haider and Sundin (2010) analyse the digital global platform of Wikipedia. Key features of this open content encyclopaedia include collective collaboration by non-experts who work to produce articles. As detailed in Jemielniak's (2014) extensive ethnography of Wikipedia, the aim is to find consensus, but the process often involves elaborate and ongoing conflict and disputation. For Haider and Sundin the online encyclopaedia overlaps with Foucault's description of libraries and museums, the idea of producing an overall archive of all times, forms and tastes in one place. Wikipedia is viewed as a continuation of the Enlightenment ideal, a duplication of traditional heterochronias that accumulate time indefinitely but in a new participatory digital or networked configuration. It is stable as a stored archive and at the same time in a fluid state, adapting and growing through constant tenacious editing. Essentially for Haider and Sundin, Wikipedia becomes a placeless place on the internet, contesting customary experiences of time and space (2010, 12).

There are clearly some heterotopic features of Wikipedia, emerging as a global memory bank, an indefinite accumulation of knowledge that can be accessed immediately. But are these heterotopic features diluted or strengthened? My contention is that the former case is most compelling. You can access Wikipedia freely from almost anywhere, dipping in and out. As part of our digital environment, along with many other online sites such as Facebook, it has become part of our everyday routine, or as Castells anticipated "part of the fabric of our lives" (2002, 1). Can something that is so close, accessible, open-ended and familiar be thought of as an example of a heterotopia which, despite these sites' eclectic family resemblances, were all primarily defined by Foucault as segregated, closed and unfamiliar? As we scroll through Wikipedia whether distractedly or with serious purpose, we are participating in a ubiquitous digital environment. This is not to dismiss significant implications of the shift in the storage of and access to information, but as I go on to argue, these shifts themselves and other transformations of our wider milieu may offer a more fruitful approach to reimagining Foucault's notion in the digital age. The actual "site" of Wikipedia dissolves amongst a new pervasive space of communication technologies.

Social media sites with greater participation and a wider sense of community have also been interpreted as new forms of heterotopia. Rymarczuk and Derksen (2014) argue that Facebook is a bounded virtual space with its own geography in terms of landscape and locations, involving a range of heterotopic qualities. Within the site, various tensions and contradictions arise; for example, users have a sense of control over social connections and yet a lack of privacy, isolating and exposing them at the same time, and she or he is unconstrained by distance and yet confined to a screen. What is deemed public and private becomes blurred, as do the boundaries between work, home and leisure. The authors note procedures for registering and "entering" Facebook and difficulties to exit, especially in deleting personal data permanently. The site also has aspects of heterochronias, recording and storing memories in text and images which can be instantly made present, as when prompted about an image posted some years previously. An ever-present timeline reveals a user's entire digital history. Moreover, the authors note that millions of users' accounts remain active after they have died, which for some offers a sort of virtual cemetery, collapsing past, present and a digital after-life (Rymarczuk and Derksen 2014, 6).

These are significant comparisons, but again we need to at least acknowledge some contrary trends. For example, Facebook's *raison d'être* is essentially about inclusiveness, openness and connectedness. It works to be participatory and to bridge temporal gaps and geographic boundaries. Although Mark Zuckerberg's ideal of "frictionless sharing" in practice often generates polarised opinion and dangerous tribalism, there is a drive towards involving everybody, everything, all of the time (Simanowski 2018, 16). The direction of travel is very different to Foucault's interpretation of heterotopias. This in itself is not problematic, except the authors attempt such a comparative analysis. As I go on to argue, in a sense we have to forget Foucault, or at least bracket off his specific principles and examples of heterotopia and take a wider perspective of our digital era. The contrary trends themselves of openness, connectivity and immediacy produce a new, different space.

Our digital environment

In exploring how the concept of heterotopia can help to understand a world that is becoming increasingly populated with digital communication devices, it is worth initially clarifying what is meant or implied by the term "digital." The term is commonly used to describe electronic technology and devices like computers and smartphones that process and store data through a binary system of ones and zeroes. However, the digital can also be understood more broadly. For example, Galloway interprets the digital in a much wider context as a theoretical concept involving any basic "distinctions" that are used to understand the world (2014, 12). In his critical celebration of François Laruelle's "anti-digital philosophy," Galloway includes in his definition of the digital any practice that defines distinct elements, from the letters of the alphabet to genetic codes. Provocatively Galloway goes further to argue that most philosophy and digitality are the same thing, as the former is above all the practice of drawing distinctions, discriminating between different elements or discrete points. In the alternative analogue domain, the world remains "unaligned, idiosyncratic, singular" (103).

Both Galloway and Serres persistently aim to unsettle dualistic formulations and philosophy's restrictive forms of reasoning, and yet they follow different paths when thinking about computer technology overall. The distinction between the two outlooks is instructive and worth a short detour. For Galloway computer technology seems to have intensified digital domains at the expense of the analogue, considering the computer itself as a practice or "a set of executions or actions in relation to a world" (2012, 13). Although we might think computer technology is an enabling device for humans to examine and to comprehend the world, for Galloway it is actually an autonomous activity in itself. The computer is a "simulating" machine, or even a "being," for storing, transmitting, processing and above all, modelling "new worlds." The premise of digital technology like computers is to interpret or grasp the world, but all it can do is manipulate, or model some artificial version of the world and in the process actually "erase" it. It is a complex machine for making ever more finer "distinctions" which removes the "real," the chaotic analogue domain (Galloway 2014, 103).

Serres also envisages a variety of capabilities of computers, including assisting or even replacing processes of the brain. But his optimistic vision of the importance of advanced digital technology radically diverges from Galloway. This can be demonstrated by Serres' striking argument that "artificial intelligence" goes back at least three millennia and "is older than intelligence itself" (2017b, 156). In exploring the ancient origins of geometry, he refers specifically to placing a stick in the ground vertically to cast a shadow. In Greek "*gnomon*" the vertical part of a sundial refers to the "one that knows or examines." For Serres, this ancient use of the perpendicular signifies "intelligence and artificial object at the same time" (154). The stick, the sun and the shadow assist the observer and form intelligence. The significance and breadth of this argument are impossible to capture in a few words but in the context of reimagining heterotopia, it is useful to compare Serres' and Galloway's very different positions here. The latter claims:

> in order to be in a relation with the world informatically, one must erase the world, subjecting it to various forms of manipulation, pre-emption, model-ling, and synthetic transformation.
>
> (Galloway 2012, 13)

In contrast, Serres argues that we have always been in relation with the world, as it were, "informatically." Technology does not erase the world; it is part of the world and has always provided a way for us to think. Put another way, Serres would contest the division between analogue and digital, seeing this as a process of dualistic thinking itself in terms of either one or the other. As discussed in the introduction, Serres persistently insists on a mixed, inclusive "third" perspective (2018, 22), a middle, as seen in his interpretation of the "*gnomon*" that shatters our division of the subject and object. In sum, Galloway argues that interfacing "effects" are ontological processes of simulation that erase the world, the ana-logue, whereas Serres considers interfacing with technology as a feature funda-mental to all human thought, a process of stimulation, rather than simulation. Technology from the sundial onwards is a gift of the world that we share. This is important because it helps to demonstrate how Serres can embrace the potential of digital technology whilst at the same time persistently undermine binary thought through initiating the disruptive inclusive third, an utterly unsettling process that I will go on to link to an alternative, dynamic conception of heterotopia.

From another perspective, Hui, in his exploration of the "existence of digital objects" (2016), also traces the intimate, overlapping tie between humans and technology. Working through a dialogue between Simondon and Heidegger, he convincingly argues that it is impossible to separate the digital from any human activity. Where distinctions arise depends on what we are trying to observe and how, the position or perspective we take, or different "levels of magnitude." For example, it is a mistake to examine electrons, or logics, in isolation from the factors that give rise to perception and intelligibility and vice versa. For Hui, the digital object is the computer screen operated by a mouse or a smartphone which we engage with through moving and scrolling. The digital "informs the user of its existence via its appearance" through familiar objects such as different forms of mobile technologies (Hui 2016, 110–11). In Heideggerian terms, these devices are "ready-at-hand." As a specialist, we might observe the computer or digital objects "as if" they are in space such as a web or network, but when we use them they are part of our inclusive and transformative spatial environment. Hui explains that we not only use digital devices at any place and time, they also occupy and transform our place, time, body and imagination. Similarly, Castells demonstrates how we not only live, work and play but also "process our creation of meaning" through technology (2002, 235). For Castells a new urban space of mobility emerges, involving complex flows of information. As Rose confirms in an extensive review of recent geographical scholarship, "digital technologies are creating new forms of urban space: sentient, circulatory and splintering" (2017, 780).

Specifically, Hui puts forward the idea of "tertiary protention" (2016, 221) – a term adapted from Husserl and also reinterpreted by Stiegler – which refers to advanced technologies that become a significant function of our physical and mental orientation. Our orientation becomes increasingly an algorithmic process which guides our immediate future – anticipating, predicting, suggesting and prompting – and is not just a matter of storage or "retentions." We can think of our increasing reliance on satellite navigation systems, or the prevalent use of mobile apps to find near-by locations, or more concerningly the rapid emergence of what Prassl describes as the "digital intermediation of work" through the local gig economy (2018, 13). Deliveroo, as an example, employ an easy supply or "crowd" of workers to deliver food to the door, with their labour, pay, place, time and direction controlled intimately by algorithms via a mobile app. As a result of this process, the deliverer of food becomes merely part of "computational infrastructure." More widely in a debate with Stiegler, Rouvroy argues that through advances in "algorithmic governmentality" we are caught in a network where each individual is merely an "aggregate of exploitable data on an industrial scale" (Rouvroy and Stiegler 2016, 9). Deleuze in some ways anticipated these developments in what he calls "societies of control" arguing that individuals have become "*dividuals*," a population of "samples, data, markets, or *banks*" (1992, 5–6 – original emphases).

There are undoubtedly many abusive and controlling elements of advanced digital technology. Stiegler (2010) describes the cultural and political dangers of what he calls "psychotechnologies" driven by global marketing industries aimed at capturing young people's minds. He argues that the result is an accumulative disordering of young people's critical attention. Stiegler does not dismiss advanced communication devices but urges us to transform them into "technologies of collective intelligence" through a thorough rethinking of education (2010, 58–59). As explored later, Serres agrees with this last point but contrary to Stiegler insists that young people have the capacity to lead in the transformation of learning and is rather contemptuous of older people who monotonously complain that "it was better before" (2017c). These debates suggest that the destructive, divisive and exploitative effects of digital technologies are in constant tension and struggle with the productive, unifying and inventive effects, but all the aforementioned agree that we are in the midst of a significant historical shift that is starting to have profound consequences on every aspect of life. We are starting to inhabit a different world.

The question remains in what way the concept of heterotopia needs to be rethought in order to convincingly capture any aspect of the complex, ever-changing and open-ended digital environment that has recently evolved and through which we are immersed. Applying phenomenological commentaries of place in Heidegger, Maggini (2017) specifically refers to heterotopia in arguing that digital environments are both an everyday lived experience and a challenge to it. When we use the internet, for example, we are "opening up a world through difference" contesting the boundaries between the customary and the unfamiliar (Maggini 2017, 470). He suggests that such an experience is in diverse ways disturbing and uncanny. This is a distinct argument regarding heterotopia and the digital domain as it focuses on having to process the challenge of a new type of dynamic experience of "place making," a formal classification of a distinctive *topos*. Focussing on the fuzzy boundary of interacting with technology in this way becomes more significant if we consider the multiple modes through which we are starting to interface with DCTs. We no longer have to sit in front of a computer to search the internet; we have an assortment of handy, mobile devices and smart speakers that provide access to the internet and various devices of orientation at any time and any place. Serres points out that in French *maintenant* [now] refers to holding in one's hand [*tenir en main*]. With mobile communication technologies, the immediate, the present is in our hands, as is the world, the global, changing our relations with time, space and others (Serres 2019, 204–8).

Taking forward Maggini's argument that focuses on the process of interfacing but avoiding any dualist notion of the everyday and the unfamiliar, our present era can be conceived as a profoundly different space, heterotopia, with multiple unsettling effects. Serres provides further signposts for exploring these effects through the overlap between his methods of enquiry and his ideas about the processes of time and space. He describes his approach to philosophy as "topological," in terms of the "science of proximities" that brings together the "most disparate things" and forms new contiguities (Serres 2018, 70). As indicated, his approach is utterly inclusive, drawing from ancient myths, literature, mathematics and the hard and soft sciences. Clashes, ties and combinations of these threads of thought open up astonishing discoveries and intuitions. Serres explains his approach by likening it to working with a sack that can be folded in different ways and placed easily inside or mixed with other sacks, or a crumpled piece of paper with creases, folds and pleats that produce unanticipated lines of association (2018, 62). He also sees time, both as we experience it and in terms of history, not in a simple, linear or progressive manner but flowing like a shifting river, dragging bits and pieces along, tangled, filtering and varying with the fluctuations of the weather. A favoured term to capture this movement is "percolation," a process of sieving (215).

Significantly for this chapter, he also suggests that a method based on topological space and the percolation of time provides a way of thinking that reverberates with the potentialities of advanced technology in our contemporary global era:

> a new space of new transports is installed on a global Earth, a space more mixed than pure, more blended, variegated, tiger-striped, zebra-striped, more multiple and inter-connected than smooth or homogenous.
>
> (2017b, 211)

He not only thinks through "mixed places, with contingent, unexpected proximities, a heterogeneous space" (2015b, 160), he also proposes that we are now
living through such a divergent space. Serres captures this well in arguing that
metaphors relating to networks or webs are misleading in suggesting a maze of set
routes, wires and nodes. We should think instead in terms of variable and revers-
ible flows of waterways or winds of different intensities and directions (2018,
215, 2019, 181–82). Concentrating on the axes of space in this emerging era takes
us right back to the beginning of Foucault's lecture and his description of the
epoch of "simultaneity," "juxtaposition," the "near and far," the "side-by-side"
and the "dispersed" (1986, 22). Foucault presciently describes this as a contemporary space of "*emplacement*" which he defines as involving mixed relations of
"proximity" involving new forms of technology, "machines," for placing, storing,
processing and distributing data. In 1967 this is a remarkable anticipation of our
present digital environment, a new, relational and utterly different space that is
variously unsettling: heterotopia. As argued previously, this different space inte-
grates instant access to Wikipedia, Facebook, Google and a host of other sites,
reversing rather than imitating Foucault's initial principles. As the rest of the
chapter explores, we inhabit this global space differently and it opens up opportunities to think and act differently.

Heterotopia as reimagining the global

We are living in an emerging global heterotopia which disturbs human existence
for better and worse. The remaining part this chapter will address the question
of how we might challenge the latter, the divisive and damaging forces of this
new global space, with specific attention to the climate crisis and the potential
destruction of the planet. As a springboard, I turn to Foucault's brief definition of
textual heterotopia in the preface to *The Order of Things* via Borges' now famous
bewildering taxonomy of animals, which for Foucault quite clearly is a heterotopia
that can occur only in language (1970, xv). Serendipitously, Serres was Foucault's
student and then colleague and helped in the development of the book, as well as
the writing of a short essay on Foucault's spatial method of study (Serres 1969).
I focus less on what Borges taxonomy is and more on what the text does. It causes
Foucault to explode with laughter, as it ruptures all the familiar ways of ordering
thought, of placing words and things together. Heterotopia here is performative
and provocative, a raw audible response, a release, a beginning. Distinguishing
between "resemblance" and "similitude," Foucault also finds this utter disorientation in his short essays on Magritte's paintings (Foucault 2008, 46). Resemblance
"refers" to something, a model of what is seen or known through a process of classification of this and that. In contrast, similitude in Magritte's paintings presents
juxtapositions of images that attack this process, affirming rather than prescribing,
and initiating a "play of transferences that run, proliferate, propagate" (49). Borges'
text and Magritte's paintings interrupt and disrupt jokingly and adventurously.

For Foucault heterotopia is an inspiration to turn back and analyse customary modes of modern thought, investigating the underlying codes that order the

human sciences. My proposal is that Serres intuits thought from such an explosive, disruptive noise but travels in the opposite direction, on the waves of Foucault's laughter, as it were, and untiringly aims to "undo the orderings of the sciences" (Serres 2015a, 40). Serres does not study what distinguishes modern forms of reason; he attempts to hunt for a new form of reasoning. Appositely laughter is at the heart of Hermes, Serres' favoured guiding figure for capturing playful inventiveness, the messenger of the gods and the transgressor of boundaries (1969, 245). Teasingly Borges' fable also upsets thoroughly thought's boundaries and connections, leaving space for a new type of knowledge to be invented. Serres' philosophy demonstrates above all that "in the places where thought has not yet been, it is difficult to think" (2015b, 54). Disruption is necessary to open a space that aids this challenging task. His method is to think in a "place without a place," a space for multiple "different voices" (210). In sum, heterotopia's provocative force resonates with Serresian endeavours to disturb our substratum of thought, break it up and experiment with different orders and relations. As Foucault puts it, in the context of heterotopia, "a disorder in which fragments of a large number of possible orders glitter" (1970, xvii).

As indicated previously, Serres considers that the internet and advanced information technology can assist in stimulating inventive thought, reversing common conceptions of its damaging effects. He acknowledges that the internet is full of constrictions and threats and far from free and accessible to all, but nevertheless sees potentialities emerging that offer "under a thousand cartographies, a topology that's crumpled, granulous, as mixed with local and global as cultures themselves" (Serres 2018, 96). The internet allows a changeable arrangement of knowledge, perspectives and relations rather than a "fixed position" in time and space. This is not about stepping into or navigating through a different virtual world; it is viewing the internet as a valuable open-ended resource for crumbling things up. In his terms it can become a sort of "topological" and "percolating" machine with a global reach (97). In these terms, heterotopia becomes a space for dynamically generating difference, transforming Foucault's conception and descriptions of a different space into an active space for stimulating relational and truly transdisciplinary ideas.

But this is not just a question of unsettling entrenched thought and reason. Serres' hope is that global technology's tendency to be divisive and exploitative has a contrary capacity to upset the powerful owners of knowledge in education, science, the economy and politics. As his eminent teacher Foucault explored so diversely, the relations of power and regimes of knowledge are inextricably linked. In his popular essay *Times of Crisis*, Serres asks: "Who will speak out" and engage with the profound dangers of globalisation, specifically the destruction of the earth and the increasing division of rich and poor (2014, 72)? In *Thumbelina* (2015) he provides a provocative answer: our hope must be a radical shift promoted by young people through their use of advanced communication technology. The way they can fluently and dexterously access and communicate through their smartphones offers a valuable instrument for change (Serres 2019, 205–6).

Thumbelina takes central stage in Serres' modern fable, continuously scrolling through and tapping her mobile devices with her thumbs. She thrives in the new digital age. Thumbelina is at the apex of transformations that started to accumulate in the second half of the twentieth century and are captured in Serres' idea of "hominescence." The most recent "*l'Hominescent*" is Thumbelina (Serres 2016, 198). According to Serres, Thumbelina not only inhabits a different time and space she also no longer possesses the same head or body of her parents. New forms of technology have altered or mutated her whole being in the way she writes, speaks and comprehends. He traces how historically knowledge has transformed in revolutionary ways, moving from the body of the storyteller, to techniques of writing and print until reaching a new form of objectification through computers and the internet (Serres 2015a, 21). Knowledge is today distributed everywhere rather than concentrated in books, libraries, schools and universities. He suggests such institutions are "old concentrations" from an era that no longer exists, but it is the philosopher's subsequent questioning that interests me here:

> Faced with these mutations, we no doubt needed to be inventing unimaginable novelties far outside the obsolete frameworks that still format our behaviours, our medias, and our projects. . . . Why have these innovations not taken place?
>
> (Serres 2015a, 15)

These "unimaginable novelties" have not materialised yet, but Serres urges the young to "reinvent everything" and refers to the legend of St Denis to explain how a new faculty for invention has emerged (2019, 183–84). In the legend, St Denis is beheaded in Paris and climbs a hill, later named Montmartre, with his severed head in his hands. Serres explains how for Thumbelina her mobile device has replaced her head, a jam-packed machine processing and sifting masses of data. For Serres much of the elaborate process of learning is now externalised in a new form. She holds in her hands a cognition that used to be part of her and which in turn unlocks a faculty of resourcefulness and creativity (2015a, 18–19). Serres suggests that the internet is like "sailing on an ocean of information" (2018, 130) and envisions it as part of a wider new space for thought and actions. In other words, DCT devices and the internet specifically provide a heterotopia, as a space to *make* difference, joining forces with Serres' persistent endeavour for "serendipitous intuition" (2015a, 40–43).

In later interviews, Serres (2017a, 23) confesses that some of his optimism, particularly about the power of the internet to promote a new form of democracy of knowledge, was mistaken, in light of such misuses as fake news, data mining and digital surveillance. He admits that the legend of *Thumbelina* is rather programmatic and utopian, and specifically does not sufficiently clarify the distinction between information, disinformation and knowledge (*savoir*). However, he insists on upturning popular assumptions, reinforced by Stiegler (2010), about how mobile technology harms young people's attention to learn. Serres is adamant that the young are no longer passive and obedient receptacles of knowledge. Order and classifications are needed for the practicalities of daily life, but replaying his consistent message, the "ill-sorted or disparate has virtues of its own" (Serres 2015a, 40). Young people may well annoy teachers with their chatter and guileful scrolling on their devices during lectures, but Serres interprets such behaviour as the start of a transformation of power and competence, a disruption that will eventually break up academic hierarchies and disciplinary boundaries, as well as other institutions and authorities. Taking up his pervasive idea of "noise" as a necessary and productive disruption similar to the "inclusive third" (Serres 2007, 8–9), he hears young people making "the new noise of the depths, a cacophony of clamouring voices" heralding the possibility of a profound change (2015a, 54). Such unruly noise has the effect of heterotopia as Foucault describes it, generating "a disorder in which fragments of a large number of possible orders glitter" (1970, xvii). Serres exposes the potential for a productive conception of heterotopia through young people's inventive and playful employment of mobile phones and other digitally mediated spatial practices.

Thumbelina is not in any pejorative sense a typical "teenager of today"; she is more a "quality of existence" that is found in young people as well as students, workers and a variety of citizens. She is a name for anyone who undermines the political institutions and hierarchies of "presumed competence" and experiments with new approaches released by the different technological space in which we are starting to live (2015a, 61). In this context I put forward Greta Thunberg, the celebrated teenager from Stockholm, as a poignant example of such an emerging quality. The teenager quickly became a global figure after sitting outside the Swedish parliament announcing she would not attend school unless political authorities would act radically to address our climate crisis. She questioned the value of education if it did not address fundamental questions about the future of our world. Some months after her single-minded protest, through the dynamism of communication technology, she had sparked ongoing worldwide school protests which by June 2019 involved more than 1,400 cities in more than 110 countries in a global day of strikes.

Greta Thunberg partly explains her remarkable stance and its impact from a single-mindedness and sense of difference produced from her autism, a special cognitive diversity that she says makes her see things directly and differently (Birrell 2019; Thunberg 2019, 30). Like Thumbelina, Greta Thunberg does not trust the competence of teachers and politicians. She announced through her practice and voice her own competence. Like Thumbelina she is adept at using mobile, global technology. For example, on a day of global school strikes, supporters are empowered by watching, posting and sharing videos of protests from across the globe via social networking sites like Twitter, Facebook and Instagram. She posts and tweets every day on all three sites and has millions of followers. In some ways this constant exchange of messages, images and videos by mainly young people to galvanise protest can be seen as a version of Serres' imagining a future communal passport or identity card, a sort of condensed symbol of heterotopia for the digital era, forming a blaze that would "mix the thousand and one different belongingnesses life meets with, undergoes and invents: in a completely different space" (Serres 2018, 96).

Greta Thunberg as an embodiment of Thumbelina draws together a variable identity that is potentially not tied to narrow "belongings" such as gender, class, religion, nation, regions and political parties (Serres 2015a, 7). In the context of the internet's elevation of tribalism and fundamentalism, as a young person she offers hope of a different voyage of connection, to something universal "born precisely from a difference" (Serres 2017b, xiii). In Serres' words: "So if you want to fight against said globalisation, fight instead, it seems to me, against particularism, the way of life of the most powerful" 2018, 217). Whether the protests make a lasting difference or not, it demonstrates remarkable features: the global intervention starts with a single child, a Thumbelina, and not established authorities; the protest spread spectacularly swiftly across the world through DCTs; it is a generational challenge targeted at world political leaders; it calls for systematic radical change rather than just highlighting lifestyle choices; and it demonstrates how new technologies that children have grown up with and through which they outpace adults can be used for action and change. Hierarchies are noisily upturned; children are teaching and leading adults and have a global reach. Politicians and others in authority may well find subtle ways of denying, ignoring or diluting their demands, but we can only hope that the Tumbelinas of this world intervene by inventing the inclusive global thought and tools to outmanoeuvre them. In a striking argument from one of her speeches Thunberg insists that we need a new politics and a new economics, but "that is not enough," we also "need a whole new way of thinking" (Thunberg 2019, 36).

Conclusion

How can we reimagine heterotopia in the globalised digital era? A plethora of DCTs and multiple access to the resources of the internet are opening up a fundamentally new space. Specific online sites are an integral part of this transformation, but I have argued that they often run in a contrary direction to Foucault's fundamental principles, features and axes of heterotopia which focus primarily on enclosures and separation. The direction of travel of openness, accessibility, participation and intimacy is contrary to such a depiction of heterotopia and yet contributes to the different space in which we live. What I propose is a sort of amalgamation and elaboration of the two incongruous sections of Foucault's original thesis: the beginning with his prescient remarks about an emerging network era, which he never returns to, and the final section that changes register and rather flamboyantly talks about the criss-crossing journey of a ship, exciting resources for the imagination. To develop this argument, I have frequently referred to elements of Serres' thought and methods. He goes further than any other philosopher in both describing and explaining an emerging new space and inventing innovative and challenging ways of exploring it.

How should we engage with this new space that is altering what we can do and think? Taking a cue from Foucault's response to Borges' "impossible" heterotopian text and crossing this with some of Serres' stimulating methods of philosophy, I have suggested that heterotopia can also be conceived performatively, as disruptive laughter, an explosion that splinters customary modes of thought and action. Heterotopia here is about confronting a different world differently. Using

Serres' fable of Thumbelina as a platform, I put forward Greta Thunberg and the global protests she has generated as a vibrant example of a new "capacity to intervene" and disrupt (Serres 2014, 25). Greta Thunberg, a striking example of Thumbelina, not only inhabits heterotopia, she is also able to generate heterotopia though her digital communication devices. I put forward a dual reimagining of heterotopia in the twenty-first century: an emerging digitally mediated space and a spatial tool for disruption of thought and practice.

Serres says that in the context of global politics we ceaselessly ask: "Who will win?" It is always a game with two players, a game that divides everything into left/right, East/West, old/new, for/against, good/bad and so on, but using one of his favoured ideas to dismantle customary thought and practice, he argues that we need the intervention of an "excluded third." For Serres this is the earth itself which through advances in technology we can comprehend in its entirety, a global subject. We should listen to the earth speak and allow it to intervene. The global is the answer to the destructive forces of globalisation, a global that joins together individuals in acts of protest and disruption to save humanity. As pointed out by Kumi Naidoo, the secretary general of Amnesty International, "the planet does not need saving" (2019, 31) because if we continue in the same way, the planet will be fine and heal, but we will not be here to damage it further. Naidoo puts forward Greta Thunberg – with no finances, no organisation and no support from traditional political authorities – as an example of inventive, experimental mobilisation. In the different space through which we live, I also put forward this young person with a hand-painted placard and a mobile phone, as an emancipated Thumbelina who has started to address the earth as an inclusive subject in an inclusive way, initiating a fresh, buoyant practice of globalisation.

Works cited

Birrell, Ian. 2019. "Greta Thunberg Teaches Us About Autism as Much as Climate Change." *The Guardian*, April 23.
Castells, Manuel. 2002. *The Internet Galaxy*. Oxford: Oxford University Press.
Dehaene, Michiel, and Lieven De Cauter, eds. 2008. *Heterotopia and the City: Public Space in a Postcivil Society*. Abingdon: Routledge.
Deleuze, Gilles. 1992. "Postscript on the Societies of Control." *October* 59: 3–7.
Foucault, Michel. 2008. *This is Not a Pipe*. Translated by James Harkness. Berkeley, CA: University of California Press.
———. 1986. "Of Other Spaces." *Diacritics* (16): 22–27.
———. 1970. *The Order of Things*. Andover, Hants: Tavistock.
Galloway, Alexander. 2014. *Laruelle: Against the Digital*. Minneapolis, MN: University of Minnesota Press.
———. 2012. *The Interface Effect*. Cambridge: Polity Press.
Haider, Jutta, and Olof Sundin. 2010. "Beyond the Legacy of the Enlightenment? Online Encyclopaedias as Digital Heterotopias." *First Monday* 15 (1): 1–13.
Hui, Yuk. 2016. *On the Existence of Digital Objects*. Minneapolis, MN: University of Minnesota Press.
Jemielniak, Dariusz. 2014. *An Ethnography of Wikipedia*. Stanford, CA: Stanford University Press.

Leu, Rada. 2018. "Almatourism: Fashion's Non-Places: Digital Complicity and Visual Codes." *Journal of Tourism Culture and Territorial Development* 9 (1): 115–27.

Maggini, Golfo. 2017. "Digital Virtual Places: Utopias, Atopias, Heterotopias." In *Place, Space and Hermeneutics*, edited by Bruce Janz, 465–77. Cham: Springer.

Naidoo, Kumi. 2019. "Bigger, Bolder, More Inclusive." *Amnesty International Magazine* 201: 30–32.

Piñuelas, Eddie. 2008. "Cyber-Heterotopia: Figurations of Space and Subjectivity in the Virtual Domain." *Watermark: California State University* 2: 152–69.

Prassl, Jeremias. 2018. *Humans as a Service: The Promise and Perils of Work in the Gig Economy*. Oxford: Oxford University Press.

Rose, Gillian. 2017. "Posthuman Agency in the Digitally Mediated City: Exteriorization, Individuation, Reinvention." *Annals of the American Association of Geographers* 107 (4): 779–93.

Rouvroy, Antoinette, and Bernard Stiegler. 2016. "The Digital Regime of Truth: From the Algorithmic Governmentality to a New Rule of Law." Translated by Anaïs Nony and Benoît Dillet. *La Deleuziana: Journal of Philosophy* 3: 6–29.

Rymarczuk, Robin, and Maarten Derksen. 2014. "Different Spaces: Exploring Facebook as Heterotopia." *First Monday* 19 (6): 1–12.

Serres, Michel. 2019. *Hominescence*. Translated by Randolph Burks. London: Bloomsbury.

———. 2018. *The Incandescent*. Translated by Randolph Burks. London: Bloomsbury Academic.

———. 2017a. "Ouverture." *Philosophie Magazine: Hors-Série* 39 (12): 21–24.

———. 2017b. *Geometry: The Third Book of Foundations*. Translated by Randolph Burks. London: Bloomsbury.

———. 2017c. *C'était Mieux Avant!* Paris: Manifest Le Pommier.

———. 2016. *Pantopie ou le monde de Michel Serres: Entretiens avec Martin Legros et Sven Ortoli*. Paris: Le Pommier.

———. 2015a. *Thumbelina: The Culture and Technology of Millennials*. Translated by Daniel Smith. London and New York: Rowman and Littlefield.

———. 2015b. *Rome: The First Book of Foundations*. Translated by Randolph Burks. London: Bloomsbury.

———. 2014. *Times of Crisis*. Translated by Anne-Maries Feenberg-Dibon. New York: Bloomsbury.

———. 2007. *The Parasite*. Translated by Lawrence Schehr. Minneapolis, MN: University of Minnesota Press.

———. 1995. *The Natural Contract*. Translated by Elizabeth MacArthur and William Paulson. Michigan, MI: The University of Michigan Press.

———. 1969. *Hermès I: La Communication*, 191–205. Paris: Les Éditions De Minuit.

Simanowski, Roberto. 2018. *Facebook Society: Losing Ourselves in Sharing Ourselves*. Translated by Susan Gillespie. New York: Columbia University Press.

Stiegler, Bernard. 2010. *Taking Care of Youth and the Generations*. Translated by Stephen Barker. Stanford, CA: Stanford University Press.

Thunberg, Greta. 2019. *No One Is Too Small to Make a Difference*. London: Penguin Books.

Van den Boomen, Marianne. 2019. *How Metaphors Matter in New Media: Transcoding the Digital*. Amsterdam: Amsterdam University Press.

Wark, McKenzie. 1993. "Lost in Space: Into the Digital Image Labyrinth." *Continuum* 7 (1): 140–60.

Young, Sherman. 1998. "Of Cyber Spaces: The Internet and Heterotopias." *M/C Journal* 1 (4): 1–23.

11 The prison as playground

Global scripts and heterotopic vertigo in *Prison Escape*

Hanneke Stuit

Introduction: the prison as playground

In the context of globalisation, prisons appear anomalous. Against a background of global economic flows, migration, and technologically induced connectivity, prisons seem fixed and isolated places, out-dated in their spatial separation of criminalised others from law-abiding citizens more generally. The boundaries of the prison are, however, not so easily fixed. On a global scale, surveillance systems feed into harrowing mechanisms of social sorting (Bigo 2008) that connect the prison and detention to structural inequity (Alexander 2010; Wacquant 2014; Fassin 2017). In relation to how prison systems organise time, Michael Hardt has noted:

> When you get close to prison, . . . you realize that it is not really a site of exclusion, separate from society, but rather a focal point, the site of the highest concentration of a logic of power that is generally diffused throughout the world.
>
> (Hardt 1997, 66)

In this sense, the prison is a heterotopia *par excellence*: it functions on the basis of a controlled opening and closing that partly isolates it from society, yet offers a concentrated lens on understanding logics of power that resonate globally. These logics, preoccupied as they are with "the problem of human emplacement," determine the principles of "propinquity, what type of storage, circulation, spotting, and classification of human elements" deemed appropriate to a given society (Foucault 2008 [1967], 15).

Apart from heterotopias of deviation, prisons are also exotic heterotopias (Foucault 2008, 18; Fludernik 2005, 24). In globally circulated films and series, for instance, the prison often emerges as a site of horrific excitement, on which fantasies of punishment, escape, and rugged criminality are indiscriminately projected. In what follows, I want to inquire after a third interpretation of the prison as heterotopia that is located in the tension between the prison as place and the prison as (imaginative) discourse. Foucault has famously equated this type of heterotopia with the figure of the mirror, "a sort of simultaneously mythic and real contestation of the space in which we live" (2008, 17) that offers sensations of alterity capable of questioning the social order. Getting at this alterity in the case of the prison thus involves analysing how conceptions of it are discursively produced, maintained, and questioned in symbolic and imaginary dimensions of popular culture. How do depictions of and engagements with the prison occupy themselves with criminality and imprisonment? And can such depictions help in changing the prison from a heterotopia of deviation into a site of alterity that questions the societal relations and emplacements of which it is the product?

In this chapter, I analyse real life gaming experience *Prison Escape*, which is housed in former penitentiary P.I. De Boschpoort in Breda, the Netherlands, and offers a playground through which people can get very close to prison without actually being incarcerated. In the game, which has a steep entry fee, participants are locked up with sometimes as many as 200 other players at a time. In three hours, visitors try to escape by way of role-playing and following narrative strands acted out by almost eighty performers. These actors portray guards (one of whom handles a sniffer dog), a hostile warden, a psychologist, a hairdresser, other prisoners, administrative staff, and janitors. The game relies on scripts and role-play, but also engages with the space of the former prison and the influence of stereotypical assumptions about the carceral, which players are invited to bring to the game. In doing so, it re-enacts a former heterotopia of deviation while relying on its players' engagement with the prison as exotic heterotopia of the imagination in a heterotopic space of play.

As Jennifer Turner points out, any visit to a repurposed prison is a complex spatio-temporal experience that can already be seen as a double heterotopia, "wherein we trace the subversion of an already subverted space" (2016, 103). In the prison museum, for instance, visitors experience an imagined moment "frozen in time," catching a glimpse of what the penitentiary must have looked like when in use. This game, however, differs from the "prison museum effect," because the space of the prison is used as a playground that foregoes all didactic traces. Players are immersed in their game experience rather than interpellated as "minds on legs" who need to be educated about an intended message (Welch 2015, 5). For better or worse, this prison tears down the "unspoken sacralisation" that the museum shares with the prison space (Welch 2015, 7–9), and explicitly engages players through their imagination and embodied experience of the game. This layering of the space creates a kind of heterotopic vertigo that further complicates the now "normalised" practice of prison tourism, in which disused prisons all over the world are turned into museums, bars, or hotels.

By analysing my own experience of the game, I will argue that this heterotopic vertigo is caused by several heterotopias rubbing against each other in the game. The imagination of the players, the power of the space as a former penitentiary, and the encounter with the script of the game work together to create moments of estrangement that make visible the separate layers of meaning present in the space.[1] The experience of the simultaneity of these layers is vertiginous because it jars full immersion in the game and pushes to the fore the player's own position in connecting these heterotopias of deviation, exotic heterotopias of the imaginary prison, and the temporarily transgressive heterotopias of play. Besides making clear that the heterotopic is thus not an attribute of space alone, but rather takes place as an embodied and situated doing, the concept of heterotopic vertigo in *Prison Escape* also describes sensations of estrangement that productively break down the erroneous but naturalised conception of the prison as an outside to society.

In order to show how this works in *Prison Escape*, I will first elaborate on the penitentiary's relation to heterotopias and play. Emphasising Foucault's claim that heterotopias come into existence through the virtual, I will then move on to the game's interpellation of its players' imaginations. This investigation leads through narratives and images stored in what Monika Fludernik has called the "carceral imaginary" (2005). By close reading the mugshot made of me while playing the game, I argue that it subdues players into the role of prisoners, while simultaneously interpolating them in globalised scripts of incarceration. These scripts revolve around the stock character of the criminal hero and, in their global appeal, tend to flatten out the more situated and historical discourses of incarceration at this specific site, as well as the at times uncomfortable process of game immersion.

In the second close reading, I delve deeper into my own experience of the game script, which repeated and exacerbated the mugshot's double effect of scripting players as prisoners and staging them as romanticised bandits. In subjecting me to prison routines, frisks, and group therapy, the game script, and the performers enacting it, doggedly pitted the role of the docile prisoner against my own desires of spectacular and heroic escape throughout the game. This caused an heterotopic vertigo in which I was acutely aware of, and confused by, my own position in relation to the different layers that made up the game experience. Rather than educating me about "the myth of the place," the vertigo caused a "bleed" from the game world to my own that triggered awareness of the erasing work done by my own phantasmagoric interpretations of globalised and stereotypical imaginaries surrounding the prison.

Heterotopia, play, and the penitentiary

Prison Escape is a liminal space that contains several heterotopic qualities at the same time. Housed in a former penitentiary, the game inevitably carries over resonances of heterotopias of deviation, where society places individuals "whose behaviour is deviant in relation to the mean or required norm" (Foucault 2008, 18). In this sense, the game does museum work in giving access to, and creating awareness of, a space that is usually closed off to the public (Wilson et al. 2017, 5). In the game's fixed three-hour duration, repeatable nature, and fixed location, it also reflects the fleeting atmosphere of the festival, and of the chronic heterotopias Foucault associates with the "discovery" of exoticised types of life that differ from one's own (Foucault 2008, 20). By extension, the fact that the game provides a space for role-play and "playing dress-up" (Fron et al. 2007) positions it as a refuge or escape from the political and the economical (De Cauter and Dehaene

2008, 97–98). However, as this reference to role-play already suggests, *Prison Escape* is first and foremost a game offering a heterotopic space of play (De Cauter and Dehaene 2008, 95–96).[2]

In *Prison Escape*, players act out prisoner roles but still follow a fairly rigid script. As such, it intersects the formality and goal oriented nature of more "traditional" games with qualities of Live-Action Role-Play. In the latter, immersion in a role determines the outcome of the game, the success of which is measured by what Stenros and Montala call "critical play" (2010, 25; Peterson 2012, 23–24). In its reliance on escape and other player fantasies, this spectacular and commodified use of the penitentiary seems to forego such critical possibilities. It commits a certain sacrilege in changing the troubled heterotopia of the prison from deviation, discipline, and crisis, into one of excitement and entertainment. Even though the prison is no longer in use, the carceral past lingers in it and the building's intended purpose scripts the place so heavily that the ethical injunction not to turn crime and punishment – and the social inequity involved in its daily mechanics (Alexander 2010; Fassin 2017) – into a spectacle remains.

Yet, in its playful engagement with participant fantasies, the game also challenges the "unspoken sacralisation" and social ordering associated with prison space (Foucault 2008, 16). In this sense, *Prison Escape* provides a scaffold for the heterotopia's mirroring function, which creates an experience of the space of the disused penitentiary "as a sort of simultaneously mythic and real contestation of the space in which we live" (Foucault 2008, 17). As such, heterotopias, especially in the case of the prison, create an awareness of how "spatial configurations (be they material/physical or social) . . . establish a certain order," but also offer sensations of alterity that question it (Wesselman 2013, 22). Foucault explains this sensation of alterity by metaphorically staging the heterotopia as a mirror, which can only be perceived as it passes through "the virtual." The reflection in the mirror is *imagined* as a world that lies behind it, causing the person looking in the mirror to wonder about their own position; am I over here or "in" the mirror, "over there"?

This virtual construction of heterotopia is crucial in understanding the workings of prison space. On the inside, and as is well known, the model of the panoptic prison creates the *fiction* of a potentially constant omnipresence and surveillance of prisoners' bodies (Bender 1987, 23–24; Božovič 1995, 9; Foucault 1995 [1975], 201). As John Bender has emphasised, however, this translates not just into Foucault's famous self-discipline, but also in an omniscient narration enforced on the prisoner. This narration "divests the criminal of narrative resources and designates a 'character' to be formulated" and enforces the script of an "alteration of being" that is supposedly completed when convicts are returned to society (1987, 202–3). By hiding prisoners from the public gaze, the penitentiary thus transforms "them into subjects, characters, objects of imaginative projection" *for* the public (1987, 202). As such, the prison generates virtual scripts with very tangible effects, both on the inside and on the outside.

In what follows, I will focus on the virtual scripts and imaginative projections in *Prison Escape* – fictions that depict, romanticise, and exoticise criminalised others in their societal absence. As my analysis makes clear, these scripts feature strongly in the heterotopic experience of the penitentiary through play and offer glimpses of alterity in estranging players from themselves. This estrangement is caused, not by making the game experience as "real" as possible, but by involving players' own associations with the carceral imaginary. Considering that these associations are mostly constructed in the relative *absence* of access to and narrative agency within the penitentiary in everyday life, this imaginary functions as an "exotic heterotopia" that constructs a safe and detached space for personal tinkering with cultural fantasies of crime and punishment (Fludernik 2005, 24). In the next section, I will analyse how this carceral imaginary is used and constructed in *Prison Escape* and how it becomes its own globally fluid inscription technology in the game.

The carceral imaginary and global scripts of imprisonment

The game's story of escape contains numerous narrative strands that partly overlap and intersect. Some of these centre on joining a gang or a renegade group of prisoners who dress up as guards, while others are more about rehabilitation and harmony. In these latter strands, prisoners take group therapy in order to gain the trust of the staff or even end up marrying one of the staff members. Interestingly, the script does not include Breda prison's infamous function of detaining Nazi collaborators after WWII, nor some of these men's successful escape to Germany in the 1950s. There is no mention of the fact that they were never extradited back to the Netherlands, causing heated political debate for almost 20 years (Olink 2004). Nor does the game reference the experiences of convicts actually detained there, which is a common feature of other penal tourism sites.

Instead, the game references, triggers, and reproduces plots familiar from the "neutral" realm of popular culture. The orange overalls that players are asked to wear, the performing of militaristic exercises by prisoners in the yard, and the warden's demand to be addressed with "Sir, yes, Sir" are examples of this. As these elements already suggest, the game mostly borrows conventions from prison narratives common in North American film, such as the character of the bad warden, sympathetic or perhaps corrupt guard, gang members with paisley bandanas, as well as the narrative climax of the escape itself (Jarvis 2004, 168, 170).[3] Of course, all of these references are effected in specific cultural contexts and carry meaning accordingly, yet they are consumed on a vast global scale that tends to give them a problematic universal valence. In *Prison Escape*, rather than effecting a mere forgetting of the site's embarrassing historical situatedness, however, these naturalised and globally distributed prison references also do their own cultural work in involving players bodily in the space of Breda's penitentiary.

According to Monika Fludernik, these ready-made themes and narratives of imprisonment are stored in the carceral imaginary: "[A] collection of culturally relevant images and associations that define our society's *idées reçues* about imprisonment" (2005, 16–17). In Anglophone literature from the Renaissance to the twentieth century, images in this repository have relied on a number of recurrent metaphors surrounding the prison's connections with hell, marriage, a tomb, the workings of society at large, a ship, a university or academy, and a sanctuary or refuge (Fludernik 2005, 10). Although the "constituents of this imaginary are historically diverse," Fludernik also emphasises that "they form a pool of associations that can be accessed in indiscriminate fashion" (17). Archaic aspects of the dungeon and oubliette, for instance, still influence contemporary depictions of the more sanitised and cell-based penitentiary (Fludernik 1999, 43). What Fludernik does not address, though, is that these elements, although they originate in specific historical and cultural settings, are also culturally mobile. For example, the orange overalls in *Prison Escape* are associated with prison aesthetics more generally, especially since the success of the series *Orange Is the New Black*, and do not just signal "prisonerhood" in the cultural contexts in which they are worn. As such, prison metaphors often persist through time and can easily cross linguistic, cultural, and geographical borders.

In these border crossings, prison metaphors can lose their situated valence and end up voicing the fascination prisons and criminals hold over the imagination of white, middle-class groups whose social standing means they are unlikely to end up in jail themselves (Fludernik 2005, 21). In this carceral imaginary, the criminal is a site of ambivalence between the sublime and the abject. On the one hand, criminals act out "our secret desires of freedom and transgression," while on the other hand satisfying – from a safe and cathartic distance – the law-abiding citizen's need to punish the violation of social norms and to be "reassured in their moral conformity" (Fludernik 2005, 21).

Besides a punitive desire, scholars have also noted a general turn in prison narratives in popular culture towards "the heroism of characters" (Aitken and Dixon qtd. in Turner 2013, 224) expressed through a focus on "the dangerous, violent, atomised experience of imprisonment . . . against which a 'prison innocent' can endure and then overcome the indignities of a brutal prison regime" (Kearon 2012, 6). The criminal and prisoner in this respect fulfils the role of an admirable bandit (Hobsbawm 1981), who provides the utopic desire of a "nobler and more meaningful time and place" and "defends against the narcissistic wound of our relative puniness and mortality" (Duncan 1996, 5). These romanticised renditions of criminals and prisoners certainly warrant Fludernik's conclusion that "the carceral imaginary remains a fantasy world, an exotic heterotopia that displaces our real-world emotions into the safe and apolitical realm of fiction" (Fludernik 2005, 24).

The mugshot taken during the game's narrative set-up seems to align with the underlying emotional structure of the carceral imaginary as discussed, yet also undermines it. Forming a long line and facing a photographer one after another, players are asked to hold a sign in front of their chest and look straight at the camera. When the picture is taken, it inscribes the role of criminal and prisoner onto the player's body and serves to immerse them into these roles. After the game, when the photograph is offered as a keepsake, these roles are prolonged and carried outside of the game setting. When I showed my mugshot to people in the weeks after I had played the game, most of them voluntarily suggested that I had succeeded

Figure 11.1 Mugshot of the author. Downloaded from *Prison Escape* Facebook page. Sander Erdmann, Studio Zakmes

well in looking like a "real" criminal. My Dutch-speaking friends even used the phrase "boeventronie," because of the expression on my face.[4]

Still, I did not strike this pose in an attempt to look "rough," but remember feeling a little disoriented by how seriously I was being interpellated as a felon by the check-in procedure of the game. In this way, the mugshot reifies the "capturing" and subjugating effects of ID photography (Campt 2017), but I was also simply embarrassed by being photographed (while role-playing). My discomfort must have caused me to raise my chin and smirk in an attempt to save face. This causes me to look down on the camera, translating into a defiant posture for subsequent viewers of the photograph, apparently fitting the role of the romanticised criminal. The contingent fact that I got soaked while waiting out in the rain before the game started gives my wet hair a greasy look and does nothing to improve the situation. Besides, the greyish filter exacerbates harsh lines on the face and some irregular facial skin, further underlining the roughness associated with prisons and prisoners.

In this sense, the shot's exaggerated roughness suggests a reality effect. The orange overalls, the filter, as well as the sign – which, crucially, is not a "real" police sign, but instead mimics such a sign in the heightened cultural imaginary of prison aesthetics – create a referential allusion to the real (Barthes 1986, 148). The photograph does not describe reality *an sich*, but denotes what viewers consider to be the category of the real (1986, 148). To clarify further: an actual mugshot would not look like this. Like most contemporary mugshots across the globe, Dutch suspect photographs have neutral and solid toned backgrounds without lines indicating height (Harlaar et al. 1998, 33, 37, 41, 67; Titulaer 1981, 88–90). Additionally, the framing here shows almost the full body and thus differs from generic police and ID photography as "frontal images, showing head and shoulders with no facial expression" (Mulcahy 2015, 85; Finn 2009, 2; see also Titulaer 1981). The orange jumpsuit is another anomaly. In 1983, the Netherlands became the first country in the world to abolish prison clothing (Leistra and Van Ulden 2016, 50). Prisoners in the Netherlands have not worn uniforms of any kind since that time, nor are they photographed in them when arrested or first brought in for detention.[5] Instead, the popular series *Orange is the New Black* and *Prison Break* come to mind here as obvious intertextual references, considering the ubiquity of the orange overall in the first series, and the grey, gritty visuality of the latter.[6]

Thus, in its eager "prison aesthetics," the *Prison Escape* mugshot ties playful subjugation and privileged social emplacement together, thereby reifying the mugshot's troubled history. John Tagg's analysis of late nineteenth- and early twentieth-century photographic portraiture – a technology joined at the hip with the rise of the mugshot in police practices (Finn 2009) – is insightful on this point:

> The portrait is . . . a sign whose purpose is both the description of an individual and the inscription of social identity.
>
> (Tagg 1988, 37)

Initially a privilege enjoyed by the higher classes, portraiture rapidly spread amongst the rising middle and lower-middle classes as it became more efficient and affordable to manufacture. Mugshots, on the other hand, signalled the technology's switch from privilege to social burden. As an inverse to portraiture, mugshots helped establish the lower classes it depicted, including criminals, the sick, women, colonised people, and the working class (Tagg 1988, 59; Mulcahy 2015, 80). To this day, although people whose mugshot is taken might never have committed a crime, they are still associated with the social undesirability, inequality, and assumed criminality inherent in the genre of the mugshot.

In its playful celebration of the genre, the game's mugshot repeats and reinforces these associations between mugshot, surveillance, and criminality with a spectacular twist. The rows and rows of photos published on *Prison Escape*'s Facebook page after each game actually call to mind the nineteenth-century rogue galleries – albums kept by police departments in which photos of people having come into contact with law enforcement were kept for future reference. In this case, viewers look for their own picture, or perhaps those of their friends, in order to share and further admire them on social media. In this sense, the photo also echoes the significant popular cultural appeal, ready accessibility, and online visibility of remarkable, collectible, or celebrity mugshots. It thus acts out a privileged social inscription of the player, signalled by their ability and willingness to engage with the topoi of the game's carceral imaginary. In this keepsake, the player has now become the criminal/hero, whose stylised abjectness can be admired and celebrated as a sign of their own adventurous – edgy – selfhood.

By making use of the aesthetic conventions, history, and popularity of the mugshot qua genre, the game ambiguously interpolates players in the globalised scripts of the carceral imaginary. The player's position in this script is clearly celebrated in the mugshot, which incorporates the social inequity at the basis of its technology and offers it up for spectacular viewing. In this sense, the mugshot should be seen as an amalgam of globalised tropes enforcing an appealing script of the criminal as hero that literally overwrites the uncomfortable situatedness of P.I. De Boschpoort. Historically speaking, this situatedness centres on the infamous escape of the Nazi collaborators or with the prison as a heterotopia of deviation and place of suffering isolated from the rest of society. In the present of the game, the situatedness foreshadowed the heterotopic vertigo I would experience later and revolved around my own discomfort about being interpellated and photographed as a prisoner.

This disjunction between the mugshot's propensity to capture and unsettle the photographed subject and its popular cultural staging of the criminal hero was brought to a head in my attempts at escape during the game. The romanticised bandit set up by the mugshot at the start of the game kept resurfacing in "free play" and constantly clashed with my interpellation as a docile and compliant prisoner by the actors and game script. The jumps and clashes between the two roles, and the "bleeds" that this caused between game world and the enclosing power of the prison setting, unexpectedly offered an alienating experience of play that served to disrupt the spectacular mechanisms at work in the game's carceral imaginary.

Heterotopic vertigo: compliance and spaces of play

From the moment they pass through the prison gates, players are forced into a passive role. The group of approximately 200 people is requested to leave their personal belongings – like phones, bags, and jewellery – at the check-in point. We are handed overalls, mugshots are taken, and everybody is fingerprinted. The guards glare at you when you walk by. We are made to perform militaristic exercises like push-ups and lectured by the warden during a prolonged lineup in the courtyard of the dome. The sound of shouting guards in the panopticon's dome is overwhelming; it causes an uncomfortable pressure on the eardrums, but involuntary physical responses, like covering your ears are not allowed. Unless instructed otherwise, players are to keep their hands along their sides at all times and look straight ahead. After a while, all players are brought to their cells.

On closer inspection of the cell, it turns out that some parts of it are "fixtures," like the on/off switch on the built-in light above the desk, while some are not. These small objects are hidden and scattered around, much like pieces on a game board, and distinguish themselves by the suggestion that they might come in handy later on. When I am in the cell, I find a poker chip, some pills, something that looks like weed but smells like oregano, a map of the prison, a letter from a former prisoner, and a flyer advertising group therapy. My cell mate and I divide the loot: she takes the letter, I take the flyer. Later, when the cells are opened for recreation, I become aware that the objects are in fact narrative clues or triggers; handing over the right object to the right person will kick-start "my" narrative in the orchestrated game script.

When the cells open, I walk around, looking for clues in the yard that will hasten my escape and allow me to win the game. The flyer suggesting group therapy to help me cope with being in prison looks uninteresting. Suddenly, amongst the crowd, I see someone who looks like a gang member. She is in black instead of the orange and dark blue overalls the other prisoners have on, she wears a red paisley bandana around her head, and has an impressive swagger. I walk up to her, but just as I am about to offer some of my contraband as a sign of good will, the sniffer dog is coming our way and our conversation is broken up by the guards. I am placed with my back against a wall and my pockets are searched. The guard finds all my contraband but puts half back in my pocket and hands the other half to the warden, who towers over me and asks me what my crime is. I say I am innocent. The woman with the bandana, who is standing next to me and is also being frisked, sniggers with contempt. "That's what they all say, curly head," says the warden. "Don't let me catch you with this mess again and get out of my sight." When I have the chance to look around, the girl with the bandana is gone and my therapy session is about to start. Perhaps getting to another part of the building, away from the domed courtyard where I am constantly at the mercy of the guards, will turn out to be useful after all.

The therapy takes place in a room in the administrative wing. There are six of us and the therapist. He gives us a piece of clay and we all have to sculpt how we feel. The other players start eagerly but I don't know what to do. I want to get out, but I don't know how to express that in clay. I roll the clay into a ball. When the therapist asks me about my "sculpture," I say I don't feel anything, not knowing how to deal with my annoyance about how long this session is taking while I should be escaping. I am confused about what the rules of the game are, the extent to which I should improvise, or whether I just need to play along with the therapist. In the meantime, the next person comes up with an elaborate story about how they want to improve themselves and turn into the flower they have sculpted. I notice the person across from me has also not really sculpted anything. She rolls her eyes while the flower sculptor talks about their feelings.

Together, we try to "work" the therapist, but he diverts our attempts at controlling the script by talking encouragingly about the benefits of his therapy. I become frustrated and notice with surprise that I am slumped in my chair, with my arms crossed before my chest. Although I know this is only a game and winning or losing does not matter, this involuntary physical response seems to suggest otherwise. Of course, embodied experience can never be equated to direct or unmediated knowledge of one's surroundings (Sobchack 2004, 5), but the gesture disorients me all the same and I am unsure how to proceed. After the session is over, I slouch behind the psychologist as he leads us back to the dome. On our way out, he remarks to the guard at the door that "this one is particularly crazy," pointing at me.

At this point in the game, I am not yet aware that the therapy script requires full compliance, rather than agency and improvisation. It takes me two more sessions to realise that every time I am recalcitrant, the psychologist interpellates me as crazy to the guards, who start calling me a troublemaker. My relation with the staff deteriorates to a point where my chances on escape are quickly fading. I am running out of time. I decide to gain the therapist's trust by starting a petition to save his favourite tree in the courtyard from being cut down by the warden. He takes this as proof that I have decided to heal myself. Eventually he even allows me to join the rest of the group in singing at the wedding of one of the convicts and a prison guard in the chapel next to the prison gate. I finally realise that the script is about the group collectively humouring the therapist and betraying him by bolting through the gate before the wedding has even started.

Before this dawns on me, however, I feel confused, thwarted, and curtailed. I really want to escape and win the game, and, as is evidenced by my vain attempts at finding more exciting escape routes through manipulation and gang membership, I clearly want to do so spectacularly. In order to achieve my main goal, however, it turned out that I was willing to adopt a compliant role, even though the role-playing in general, and this role in particular, made me uncomfortable. Like in the mugshot, both the demand for a docile, rehabilitated prisoner and fantasies of criminal heroism are present in my experience of the script. In fact, the experience makes clear that the one role cannot properly function without the other. The prisoner is not docile if she is not first subdued. If the prisoner is not subdued, there seems little reason for resistance.

My game experience was structured by the fact that being captured by the mugshot triggered discomfort and a performance of defiance, even before I was fully immersed in the game. Once "in it," I leaned on my own interpretation of the prisoner role, which significantly echoed the spectacularised repertoire of the globalised carceral imaginary in its universalised insistence on prison gang membership as exciting. I veered from immersion in acting out the prisoner role according to my own desires in "free play" (Sicart 2014, 51), and ruefully complying to the character that the actors allocated to me so that I could still win. However, after being scorned by both warden and criminal hero early in the game, who thus dashed my hopes at working a spectacular plot, I gradually took on the character of the subdued troublemaker.

Throughout the game, however, I also found myself wondering, not just about how to win, but about what was happening to me. What must it be like to be derided by fellow prisoners and deemed to be unfit for (naively coveted) gang membership? To be randomly frisked during recreation and scolded by a warden who picks on your physical traits? To have to sit through therapy sessions the narrative of which you find unhelpful and tiresome? To still go there because you cannot find a better way to spend your time? By formulating these questions, I do not mean to suggest that playing *Prison Escape* equals the all too real "pains of imprisonment" (Sykes 1958). What happened is that the force of the game script adjusted the role I wanted to play in the prison and prevented me from acting out my own inflections of the carceral imaginary. The script itself became like a prison, making me acutely aware of my own involuntary bodily responses to frustration, like slouching in my chair and being cross at a fake therapist. The panoptic prison in Breda may no longer be in use, but *Prison Escape* certainly pushed reflection on the experience of incarceration in general.

Writing in the context of Nordic Live-Action Role-Playing games (larp), Jaakko Stenros has argued that this friction between the game world and the real world is exactly what constitutes the power of larping. The fact that one can never fully coincide with one's character "causes friction between the everyday and the diegetic, the player and the character," and provides an "automatic distancing, a built-in alienation effect like Brecht's *Verfremdung*" (Stenros 2010, 300). Miguel Sicart, in *Play Matters*, describes such processes as bleeds, when "the transmission of experiences and knowledge from the activity of play [transfer] to our worldview" (2014, 67). Play relies on this state of limbo: you immerse in play but without losing grip on reality. It suspends reality in a way that is strongly reminiscent of the willing suspension of disbelief, but is at the same time very serious (Huizinga 1950, 8–11). This duality allows for play's ambiguous social position as being central to culture, but also partly existing outside of society's rules. According to De Cauter and Dehaene, this is precisely what makes the space of play heterotopic, because "in its formal separation from the rest of the world, [it] presents a world of instability and possibility" (De Cauter and Dehaene 2008, 96). This possibility opens up "a profoundly ambiguous terrain marking both the moment of man's imprisonment within the norms of culture and the threshold of liberation, or, more likely, temporary transgression" (2008, 96). In other words, it is precisely the fragility and instability of play that make it productive and allow for a temporal transgression of cultural norms that would otherwise not be possible.

Following De Cauter and Dehaene, play in *Prison Escape* then functions in a similar way to Fludernik's carceral imaginary. As an exotic heterotopia, *Prison Escape* showcases the catharsis of acting out criminalised desires of defiance and resistance that would be inappropriate in most cultural settings. However, as became clear from my close reading earlier, my desire to act out these stereotypical scripts was thwarted throughout the game. What caused "bleeds" to occur were in the end not moments of transgression in which I full immersed myself in playing one of the two prisoner roles. Instead, most of my game experience was characterised by a sense of vertiginous disorientation caused by the tension between the roles. As such, the game alienated me from a unified role or identity, rather than cleansing it through safe transgression. Slumping in my chair in frustration was not cathartic. I was wasting time; I was being treated like a child, forced into monolithic structures of thought. In the end, the forceful persistence of this script of compliance meant that I experienced entrapment much more than defiance. Far from running along the tiers of the dome looking for more contraband, if I ever wanted to win, I had no choice but to comply.

Conclusions

Prison Escape relies heavily on the fetish of the prison, but in doing so also forms a complex heterotopic layering that, from the materiality of the panoptic prison in Breda, shoots off in many directions. In breaking through the unspoken sacralisation of the closed-off prison, it involves globalised and exoticised carceral imaginaries in its actual space and allows the heterotopic effects of this encounter to reverberate beyond its walls after the game is over. The mugshot underlines and flaunts this multidirectionality. It utilises the double social inscription of the technologies of ID photography and portraiture as a means of immersing players in the game. Simultaneously, it captures this subjection to the prison script as a souvenir; this is what you would look like if you had been a bandit. In the ambiguous seriousness of play, the mugshot offers the player up for spectacular display as a romanticised criminal hero, and, in its tacky insistence on Barthes' reality effect, rips off the player's privileged mask of being able to engage with the prison in jest, rather than in earnest.

The game's strongest suit lies in its related potential to induce heterotopic vertigo. The vertigo arises when the scripting power of the prison as a heterotopia of deviation – in this case its accompanying social discourses of rehabilitation and docility – and the exotic imaginaries that the players project on it are registered as separate, yet cooperating layers of signification. The therapy script acted out a strong injunction to comply that laid bare the signifying layers of the prison and ultimately led to feelings of entrapment. These feelings arose, perhaps all the more strongly, because the actors pitted the injunction of the script against the fantasies of a spectacular and daring escape provided by the globalised repository of the carceral imaginary. Having to constantly negotiate these two roles is what eventually caused a bleed to occur – between the two roles, but also between the world of the game, the physical experience of it, and an awareness of the intensity of the prison discourse the game mimicked. All of this, meanwhile, was enacted in a place that may no longer function as a heterotopia of deviation, but certainly still impresses itself on the player as such.

Ultimately, then, the globalised and exotic heterotopia of the carceral imaginary and the situated heterotopia of the prison space may not be as opposed as they appear on first sight. To be sure, the particular histories of the panoptic prison in Breda fall out of the game script and are overwritten by carceral constructions that have become unmoored from the cultural contexts in which they were effected. In this sense, *Prison Escape* erases the parameters of its own locatedness and runs the risk of circulating empty signifiers of incarceration. Viewed differently, however, the game refuses to treat the local and particular as necessarily disrupted by global scripts of incarceration. Instead, the therapy script incorporated the influence of global imaginaries on the situated prison and showed how these discourses connect and clash. In this sense, the heterotopic vertigo in *Prison Escape* may serve to scrutinise the imaginaries created by cultural blindness for the penitentiary and demystify the artificial boundary between prison and society these myths keep in place.

Notes

1 Although I will touch on the effects of the building throughout this paper, the discursive and imaginative interactions with the space are emphasised. For an analysis of how the game utilises the panoptic effects of the building to trigger player escape fantasies, see Stuit 2020.
2 In this chapter, I refer to play in general, taking into account the fact that play takes place on a continuum between highly formalised games with "tedious conventions" (ludic) and free forms of play that involve higher levels of improvisation (paida) (Caillois 1958, 13). In terms of literal space, a play space is less formal than a game space, but the two can intersect. A game is a space for play with its own logic, materiality, and rules (Begy 2017, 722), but a play space is not always located in a game. Play space is more susceptible to what Sicart calls "free play," which is still determined by the space and context, but not by any rules (2014, 51).
3 The focus on gangs, particularly – which are much less prevalent in Dutch prisons than in some other countries – seems to suggest a globalised imaginary.
4 The Dutch word *boeventronie* or the less regional *boevenkop* literally means "crook's mug" or "gallow's face."
5 The photographs are called mugshots in the game's discourse, but considering that they are taken while being "checked-in" to the prison (rather than when arrested) in the game world, they are actually prisoner photos. Mugshots are taken when suspects are arrested and taken to the station.
6 Both series are readily accessible via Netflix in the Netherlands at the time of writing.

Works cited

Alexander, Michelle. 2010. *The New Jim Crow. Mass Incarceration in the Age of Color-blindness*. New York and London: The New Press.

Barthes, Roland. 1986. "The Reality Effect." In *The Rustle of Language*, translated by Richard Howard, 141–48. Berkeley and Los Angeles: University of California Press.

Begy, Jason. 2017. "Board Games and the Construction of Cultural Memory." *Games and Culture* 12 (7–8): 718–38.

Bender, John. 1987. *Imagining the Penitentiary: Fiction and Architecture of Mind in Eighteenth Century England*. Chicago and London: University of Chicago Press.

Bigo, Didier. 2008. "Globalized (In)Security. The Field and the Ban-opticon." In *Terror, Insecurity and Liberty. Illiberal Practices of Liberal Regimes After 9/11*, edited by Didier Bigo and Anastassia Tsoukala, 10–48. New York: Routledge.

Božovič, Miran. 1995. "Introduction: An Utterly Dark Spot." In *Jeremy Bentham: The Panoptic Writings*, edited by Miran Božovič, 1–28. London and New York: Verso.

Caillois, Roger. 1958 [2001]. *Man, Play and Games*. Translated by Meyer Barash. Urbana and Chicago: University of Illinois Press.

Campt, Tina M. 2017. *Listening to Images*. Durham and London: Duke University Press.

De Cauter, Lieven, and Michiel Dehaene. 2008. "The Space of Play. Towards a General Theory of Heterotopia." In *Heterotopia in the City: Public Space in Postcivil Society*, edited by Michiel Dehaene and Lieven De Cauter, 87–102. Abingdon: Routledge.

Duncan, Martha Grace. 1996. *Romantic Outlaws, Beloved Prisons. The Unconscious Meanings of Crime and Punishment*. New York: New York University Press.

Fassin, Didier. 2017. *Prison Worlds: An Ethnography of the Carceral Condition*. Translated by Rachel Gomme. Cambridge and Malden: Polity Press.

Finn, Jonathan. 2009. *Capturing the Criminal Image: From Mug Shot to Surveillance Society*. Minneapolis and London: University of Minnesota Press.

Fludernik, Monika. 2005. "Metaphoric (Im)prison(ment) and the Constitution of a Carceral Imaginary." *Anglia. Journal of English Philology* 123 (1): 1–25.

———. 1999. "Carceral Topography: Spatiality, Liminality and Corporality in the Literary Prison." *Textual Practice* 13 (1): 43–77.

Foucault, Michel. 2008 [1967]. "Of Other Spaces." Translated by Lieven De Cauter and Michiel Dehaene. In *Heterotopia in the City: Public Space in Postcivil Society*, edited by Michiel Dehaene and Lieven De Cauter, 13–29. Abingdon: Routledge.

———. 1995 [1975]. *Discipline and Punish: The Birth of the Prison*. New York: Vintage.

Fron, Janine, Tracy Fullerton, Jacquelyn Ford Morie, and Celia Pearce. 2007. "Playing Dress-Up: Costumes, Roleplay and Imagination." *Philosophy of Computer Games* (24–27): 1–23.

Hardt, Michael. 1997. "Prison Time." *Yale French Studies* (91): 64–79.

Harlaar, Martin, Richard Hengeveld, Jan Pieter Koster, and Anne Roos. 1998. *Silent Witnesses: Photographs from the Amsterdam Police Archives*. Bussum: Uitgeverij Toth.

Hobsbawm, E. J. 1981. *Bandits*. New York: Pantheon books.

Huizinga, Johan. 1950 [2014]. *Homo Ludens: A Study of the Play Element in Culture*. Mansfield Centre: Martino Publishing.

Jarvis, Brian. 2004. *Cruel and Unusual: Punishment and US Culture*. London and Sterling, VA: Pluto Press.

Kearon, Tony. 2012. "Alternative Representations of the Prison and Imprisonment – Comparing Dominant Narratives in the News Media and in Popular Fictional Texts." *Prison Service Journal* (199): 4–9.

Leistra, Gerlof, and Annemarie Van Ulden. 2016. *Biografie van de Bajes. De Roerige Geschiedenis van de Bijlmerbajes*. Meppel: Just Publishers.

Mulcahy, Linda. 2015. "Docile Suffragettes? Resistance to Police Photography and the Possibility of Object-Subject Transformation." *Feminist Legal Studies* (23): 79–99.

Olink, Hans. 2004. "De Zeven van Breda." *Nrc.nl*, February 7, 2004.

Peterson, Jon. 2012. *Playing at the World: A History of Simulating Wars, People and Fantastic Adventures from Chess to Role-Playing Games*. San Diego: Unreason Press.

Sicart, Miguel. 2014. *Play Matters*. London and Cambridge: The MIT Press.

Sobchack, Vivian. 2004. *Carnal Thoughts: Embodiment and Moving Image Culture*. Berkeley, Los Angeles, and London: University of California Press.

Stenros, Jaakko. 2010. "Nordic Larp as Theatre, Art and Gaming." In *Nordic Larp*, edited by Jaakko Stenros and Markus Montala, 300–15. Stockholm: Fëa Livia.

Stenros Jaakko, and Markus Montala. 2010. "The Paradox of Nordic Larp Culture." In *Nordic Larp*, edited by Jaakko Stenros and Markus Montala, 12–29. Stockholm: Fëa Livia.

Stuit, Hanneke. 2020. "Carceral Projections: The Cell and the Heterotopia of Play in *Prison Escape*." In *The Prison Cell: Embodied and Everyday Spaces of Incarceration*, edited by Jennifer Turner and Victoria Knight. London and New York: Palgrave Macmillan.

Sykes, Gresham M. 1958. "The Pains of Imprisonment." In *The Society of Captives: A Study of Maximum Security Prison*, 63–83. Princeton, NJ: Princeton University Press.

Tagg, John. 1988. *The Burden of Representation. Essays on Photographies and Histories*. New York and London: Palgrave Macmillan.

Titulaer, Chriet. 1981. *Fotografie bij de Politie*. Deventer and Antwerp: Kluwer.

Turner, Jennifer. 2016. *The Prison Boundary – Between Society and Carceral Space*. London: Palgrave Macmillan.

———. 2013. "The Politics of Carceral Spectacle: Televising Prison Life." In *Carceral Spaces: Mobility and Agency in Imprisonment and Migrant Detention*, edited by Dominique Moran, Nick Gill, and Deirdre Conlon, 219–47. Farnham am Main and Burlington: Ashgate.

Wacquant, Loïc. 2014. "Marginality, Ethnicity and Penality in the Neo-Liberal City: An Analytic Cartography." *Ethnic and Racial Studies* 37 (10): 1687–711.

Welch, Michael. 2015. *Escape to Prison: Penal Tourism and the Pull of Punishment*. Oakland: University of California Press.

Wesselman, Daan. 2013. "The High Line, 'The Balloon,' and Heterotopia." *Space and Culture* 16 (1): 16–27.

Wilson, Jacqueline Z., Sarah Hodgkinson, Justin Piché, and Kevin Walby. 2017. *Palgrave Handbook of Prison Tourism*. London and New York: Palgrave Macmillan.

12 Dramatic heterotopia

The participatory spectacle of Burning Man

Graham St John

Introduction

Reassembled annually in Nevada's Black Rock Desert, Burning Man is a large-scale fire-arts gathering widely recognised as a world leader in inclusive and participatory events.[1] Otherwise known as Black Rock City (total population 80,000 in 2018), Burning Man attracts a predominantly white middle-class populace. While its inclusive agenda is debatable, Black Rock City (BRC) is notable for the diversity of *habitués* populating its festal precincts. Equally noteworthy is the event's dynamic ethos – the Ten Principles[2] – the ongoing performance of which ensures the survival of this heterogeneous frontier event. BRC is understood to possess contradictory and perplexing features that are integral to its operational dynamics (Pike 2000; Davis 2005; Gilmore 2010), and is reckoned an enclave of complimentary paradox (Rohrmeier and Starrs 2014). With BRC as a case study, the chapter adapts Foucault's "heterotopia" to further understand this complex transformational "other space." It addresses the performance of paradox native to other spaces and spaces of otherness – i.e. contested sites persistently dogged by internal contradictions. Adopting the concept of *dramatic heterotopia*, it illustrates how paradox, not to be conflated with hypocrisy or "false consciousness," is the dynamic currency of participatory event-cultures.

The chapter traces the exchange between two archetypal event figures: the *artist* and *tourist*. While the former is the virtuous participant whose gifting practices, voluntary service, and labours of gratitude are integral to re-creating BRC, the latter connotes the spectator, mythologised as a ravenous, even parasitical, consumer of packaged entertainment. There are four sections. The first introduces BRC as a *hyperliminal* "other space" – a frontier carnival resettled seasonally and occasioning heterogeneous othering. The second positions this research in the nascent field of transformational event studies, interpolating BRC's stature as a *participatory spectacle*, the mode of festal being in which "burners" make a "spectacle" of themselves. The third section introduces the Ten Principles, a threatened ethos integral to Burning Man. The fourth explores Burning Man's role as a *redressive artopia*, an idea informed by an understanding that heterotopias are stages for the performance of contradictions in the wider society. According to Dehaene and De Cauter: "Heterotopias are aporetic spaces that reveal or represent something about the society in which they reside through the way in which they incorporate and stage the very contradictions that this society produces but is unable to resolve."[3] Working across artistic media, the burner community has responded to an insipid culture of convenience undermining the event's ethos. This crisis exposes a long-simmering tension (i.e. between "participants" and "spectators"). Turner's "social drama" model illustrates how the community negotiates this crisis in projects that dramatise paradox. Three art projects illustrate how Burning Man holds the potential to transform eventgoers and enable burner identity by navigating boundaries separating the artist/tourist, producer/consumer, participant/spectator, and self/other. Demonstrating that heterotopias possess heuristic value in the study of intentionally transformational events, this event-city case study offers a unique contribution to the heterotopology of cities.

Burning Man: an other world

Burning Man commenced in 1986, when primary founder Larry Harvey and his carpenter friend Jerry James torched an eight-foot effigy on Baker Beach, San Francisco, on the summer solstice. The ritual was repeated annually, with the expanding crowd at the unpermitted event prompting police to intervene, preventing the burning of "the Man" in 1990. That year, Harvey and friends were invited to burn their effigy in Nevada's Black Rock Desert as part of "Zone Trip #4: Bad Day at Black Rock," an event organised by the San Francisco Cacophony Society, a loose affiliation of pranksters founded by John Law in 1986 (Evans, Galbraith, and Law 2013). The weeklong event has transpired over the Labor Day holiday every year since, with the phenomenon evolving from a small remote gathering into a global movement.

While BRC recurs in an "other space," its spatial otherness is multiplex. Situated in the Black Rock Desert, the event is remote from its participants' places of residence. Additionally, it is situated on a 200-square-mile salt flat or playa (known to burners as "the playa"). Various factors characterise the playa's topographical otherness: the moisture sapping heat, the freezing overnight temperatures, the wind and dust storms that hamper visibility and play havoc with equipment, the sudden downpours converting dust into clay and restricting mobility. Such extremes compel a strong ethic of self-reliance, as embodied in the principle of *Radical Self-reliance*. Furthermore, as Fox (2002) noted, since playas are absolute deserts with features uniformly distributed around 360° – "isotropic" spaces – they are disorienting for members of a species that has evolved in savannas and forests. That such alien conditions induce cognitive dissonance suggests the playa is an exemplary *atopia* – that which Spariosu identifies as "an emotionally neutral liminal space" of opportunity, a space of "radical liminality" (Spariosu 2015, 33; St John 2019a). This observation points to another way in which the playa is spatially other: it is a performance space permitting experimentation with conduct that breaks with social conventions. From the performance of playa personas (alter-egos), and the queering of gender and sexual identities, to unconditional gifting, the playa approximates "thirdspace," a space "radically open to additional otherness" (Soja 1995, 61). It may even be consistent with "Third Space," i.e. ambivalent and indeterminate geographies of difference that disrupt oppositions like self/other and male/female, potentiating cultural hybridisation (Bhabha 1994). Recognised as a "blank canvas," the playa illustrates a unique incidence of the "radical openness" said to define performance spaces that are always provisional, unfinished, and relational (see Massey 2005). When participants refer to their "on-playa" existence, they invoke a virtual space upon which they stage alternate – burner – identities. That this space of possibility facilitates transit to alternate subjective "places" is well captured in the short film *Oh, The Places You'll Go at Burning Man!* (Saunders, Howell, and Walsh 2012), in which a motley cast of BRC denizens recite the narrative of the Dr Seuss story *Oh, The Places You'll Go*. These expressions are encouraged by the on-playa dissolution of the normative division of private and public space, especially apparent in the region removed from camps and residential areas known as "deep playa." Helpful in understanding the spatial significance of deep playa is Greek architect Hippodamus's concept of the "third sphere," which, by contrast to public and private spheres, is sacred or *hieratic* space. Neither public (political) nor private (economical) space, deep playa is *hieratic*, a qualification that is said to render "the otherness of other spaces – *les espaces autres* of Foucault – explicit" (De Cauter and Dehaene 2008, 90). The spaces of the *polis* that correspond to this third category "do not abide by the binary oppositions that stabilise the distinction between *oikos* and *agora*: exclusive versus inclusive, kinship versus citizenship, hidden versus open, private property versus public domain" (ibid, 91).

Not only is the playa a different physical space that permits difference, it accommodates diverse and sometimes incommensurate forms of difference. Consequently, the playa is a hotly contested event-space. Seasoned participants include the thousands building and maintaining event infrastructure. Experienced burners are employed or otherwise volunteer within dozens of departments, assuming a vast array of roles over the course of the event. By contrast, there is a large population of newcomers, with the rate of "virgin" entrants growing to an average of 37% for the years 2013–2017.[4] With this upsurge, the playa has attracted a category of participant with an aura of ambivalence; those seeking convenient methods of consuming the otherness of the playa and consequently growing their cultural capital. Concern is not only expressed about this "tourist" other in a field of otherness, but to those who enable, secure, and profit from their presence.

From veterans to neophytes, participants possess varied expectations, sometimes harbouring discordant views of Burning Man. The magnitude of variation is evident in the panoply of build crews, theme camps (almost 1,500,), art projects, and service providers, including those volunteering for the Department of Public Works, Lamplighters, Emergency Services, and Gate, Perimeter, and Exodus. The event harbours sharp subcultural incongruity, evident in the juxtaposed operations of the Black Rock Rangers (over 500 non-confrontational community mediators trained in dispute resolution techniques [Gómez 2013]) and Rangers of the Bureau of Land Management (BLM, of six different law enforcement agencies among a dozen state and federal agencies present), the federal body managing use of the playa. While the playa is identified as "home" or "home-like" for a significant population of burners – as revealed in census data collected annually from 2005 – it is a complicated homeland.[5]

While BRC is heterotopological, it features unique zones that are distinctly heterotopian. The playa is an ecology of "other spaces." Notable is the Temple, evolving as an event tradition after David Best led a build in 2000. What Foucault might have called a "counter emplacement," the Temple is a site where play, remembrance, and catharsis combine to form a distinct aesthetics of *otherness*. The Temple is placed at "12 o'clock" in "deep playa," thereby removed from Center Camp (downtown), and spatially other to residential areas. Within the enclosed Temple, public displays of grief are immediate, as are the litany of inscriptions, hanging memorials, artworks, and personal performances evoking warm and social approaches to bereavement, death, and mortality. As an experimental site for grief emergent within the social laboratory of the playa, the Temple is an "other space" nestled within the "other space" of BRC. Not only does conduct within this densely crowded space deviate from that at isolated sites hosting normative and constrained expressions – e.g. cemeteries – it is distinct from event conduct outside the grounds of the Temple.

Temple Burn and Burn Night, the two main rites, offer a notable contrast. On the penultimate night, Burn Night is a breathtaking fire performance in which the populace congress at the eponymous effigy to witness its fiery obliteration amid a spectacular pyro arts display. This is an occasion for ecstasy and revelry, backgrounded by the riotous noise from multitudinous "Mutant Vehicles" equipped with sound systems. The partying continues across the city and throughout the night. Temple Burn occurs on the final (Sunday) night, attracting a significant proportion of the population. Contrasted to the previous night's saturnalia, Temple Burn is a solemn ceremonial observance. As has been remarked: "If every other playa site invites irreverence and irony, this is still the one that takes meaning most seriously" (Pike 2011). This is a remarkable contrast, not least since Temple Burn occurs in an adjacent location to Burn Night, attracting many of the same participants, separated by one day. These Dionysian and solemn rituals are integral to the "festival morphology" (Falassi 1987, 6) of Burning Man.

While Burning Man's popularity is also signalled in a proliferation of media reports and academic scholarship, it defies classification. Since an effigy was first raised (and razed) on Baker Beach, the hubris accumulated as stakeholders invested material, emotional, and intellectual resources in the event. The commitment to capture and codify Burning Man was lampooned in "The Burning Man Phrase Generator," a device enabling users to arrive at a random four-word phrase, such as "pyro cultural lifestyle revolution" or "retro tribal dada orgy."[6] The manifold combination-signifiers randomly cranked out from this device – ostensibly 160k – satirise commentators believing they possess the real meaning behind Burning Man (the anthropomorphic statue is not imbued with official meaning). Revelling in the absence of authoritative narrative, this device may be the ideal Dadaist expression of heterotopology. As a playful celebration of the event's heteroclite character, the device evokes a suspicion shared among underground scenes that once their event, ritual, or subculture is named, represented, and incorporated – above all as a commodified product – it is disempowered. At the same time, the Phrase Generator implies that this "thing in the desert" is impossible to circumscribe, that it is not unidimensional, and that it escapes the noose of logic.

The futility of identifying the anomaly appears guaranteed by the widening complexity of "Burning Man," a phenomenon that is simultaneously: a unique space (the Black Rock Desert playa); a festive fire-arts gathering; a temporary city; a nonprofit organisation (the Burning Man Project); a global cultural movement (a network of Regional Events in 35+ countries); and a land steward (owner of Fly Ranch, Nevada). A complex ethical framework that derives immediately from recurrent life on the playa, the Ten Principles possess a superordinate status across all the elements listed.

This chapter limits focus to the role of BRC in this multifaceted phenomenon.[7] While BRC is not a legally incorporated city with elected officials, it has services simulating those of a city: e.g. departments, radio stations, a post office, a hospital, an airport, and bars, operated by volunteers and without monetary transactions. Services that were initially labelled with comic irony, and as a parody of official culture, including the Black Rock Rangers and the Department of Mutant Vehicles DMV), would become operational "departments" integral to a fledgling city, that while originally wrapped in ironic parentheses would grow to behave like an actual city – indeed, so much so that the Conference of Mayors sent a delegation of 11 US mayors to BRC in 2018 (Mitchell 2018).

Among the ideas most commonly deployed to apprehend BRC is Turner's concept of "liminality" (e.g. Pike 2001; Clupper 2007; Gilmore 2010). Following the meaning implicit to Van Gennep's (1960) rites of passage model, the application of this concept implies that BRC performs a structuring or ordering effect that facilitates outcomes in the post-liminal (or "default") world. That is, it is transformational. Among the consequences of this liminal experience is the desire for more liminality, as evident in the recurrent assemblage of BRC and its progeny events. But if BRC is transformational, its experimental liminality is excessive, ambivalent, and infused with paradox. Neotribe theory (Maffesoli 1996) assists our understanding of this hyperliminality. As burners possess multiple sites of belonging and identification within the event – e.g. family camp, art projects, build crews, departments, regional networks – their identities are shaped by profuse aesthetics, utopics, and conduct. When an adaptable concentric grid design enhanced spatial density in the mid-1990s (Garrett 2010), the consequential intensity of life on-playa reveals a culture as promiscuous as it is multitudinous. Rapprochement between event-tribes, dubbed "polycampory,"[8] is the legacy of a frontier event founded upon *ludic* and *civic* propensities. Heir to the Cacophony Society and their surrealist field expeditions (or Zone Trips), the desert adventure of Burning Man was, from its inception, infused with an impulse for the unpredictable. Over repeat instalments, the reproduction of chaos necessitated a form of governance. The interfacing of ludic/civic predilections has given shape to an immediate-civilisation recurring in an unforgiving natural theatre. The hard won confluence of licentiousness and pragmatism underlies the freak logic of this marginal centre: a field of possible impossibilities, a strange flowering in the void, a theatre of creative-destruction.

Escalating appeal culminated in population pressures by 1996, when attendance (8,000) doubled from the previous year. In the following year (1997), the entrance fee doubled to $65 (by 2019, approx. $500). Anxieties grew over increased media coverage and law enforcement. There were concerns about the proliferation of "frat boys," "ravers," and other "tourists." Additionally there were anxious reactions to the formation of an elite managerial class and the prospect of "selling out." That concern provoked the first annual art theme. Inspired by Dante, The Inferno poignantly dramatised the corporate takeover of Burning Man (Doherty 2006, 105). Camps formed around divergent philosophies. For protagonists, the event would continue to expand in perpetuity. Others thought the event had a used-by date. For these antagonists, BRC appeared more and more like the urbanity they were trying to escape. A divide separated managers (embodied by Harvey) seeking legitimacy from detractors (notably Law) desiring impermanence. The former vision necessitated a legal entity (i.e. Black Rock City LLC, formed in 1997), a formal staff structure with operations manuals, financial oversight, health and safety planning, insurance, dialogue with county sheriffs, and the permit-issuing BLM, as well as media liaison protocols. The latter camp required that Burning Man remain clandestine and, not unlike the TAZ (Bey 1991), vanish. While the event *would* vanish on a seasonal basis – becoming "the largest leave no trace event in the world"[9] – the conviction here was to resist formalisation, incorporation, and spectacularisation.

For Law, who left the organisation in 1996, BRC was crumbling under the weight of its own contradictions. In what he observed as a "two tier" event, "the wealthy people come out in their RVs and are serviced by the poor punks and hippies who organise the work to make the event happen" (in Silver 2000). Continuing his screed fifteen years later, Law felt the event had become the perfect vacation for the "code slaves" of Silicon Valley. "It's a Dionysian expression of controlled licentiousness" where "your rebellion is controlled in a completely and brilliantly designed and controlled box where you can do whatever you want" (in Scaruffi 2015). While freedom remains essentially illusory in this perspective, the Burning Man Project (BMP) was not devised to "exterminate freedom," according to Harvey, but extended permissions to greater numbers (in Bonin 2009). That the playa could be a "world of discipline or emancipation, resistance or sedation" (Foucault 2008, 25, n15),[10] illustrates the ambivalence native to heterotopias.

Principles and paradox

The principal means by which permission is extended is the Ten Principles. Concerned to make a reproducible ethos, these were formulated by Harvey and the Regionals Committee in 2004 and are the subject of terse debate. Navigating the principles in-situ is considered pivotal to "acculturation" to the event ethos, and integral to being a burner. A challenge manifests in commitments to practices embodying values associated with a singular or limited range of principles. For example, participants who exalt *Radical Self-expression*, which celebrates "the unique gifts of the individual," may neglect other principles like *Communal Effort*, the positive valuation of "cooperation and collaboration." Theme camps expressly committed to *Gifting* may fail to observe other principles like *Leaving No Trace*. How the community accommodates participants with disparate expectations and motivations, or whose behaviour is inconsistent, is an organisational predicament. Educating burners on the virtues of civic mindedness – as embodied in *Civic Responsibility* – has been integral the BMP's education program. For example, Project Citizenship is a cross-departmental effort launched in 2017 to address non-participation.

The Burning Man Philosophical Center[11] was earlier purposed to foment discourse interrogating the Ten Principles, with Harvey citing guidance from William James: "Belief is thought at rest." It was emphasised that the principles are not ideologies, commandments, or requests, that they "do not precede immediate experience," that they coexist in "an ecosystem," and that they are in a relationship of creative tension. "Philosophy occurs when principles collide, and we should allow these Principles to interpret and interrogate one another. Our philosophy, in other words, is muscular – it depends on the capacity of its assumptions to do work" (Harvey 2013). Rather than a sign of stagnation, for Harvey, paradox is an engine of creativity, and the principles, while conflictual, can only be understood in the lived environment of the event-space.

While BRC has been the stage for a culture war waged between surrealist and civilising tribes, and has seen ludic and civic passions encoded in *Immediacy* and *Civic Responsibility*, the playa (especially deep playa) acts as a frontier for the suspension of hostilities. An interplay of freedom and control, dialogue between ludic/civic valuations in BRC is somewhat consistent with the tension found pivotal to the "utopian idea of modern society" (Hetherington 1997, 51). As Hetherington has argued, heterotopia demonstrate processes of social ordering more than the establishment of a new order; where utopia is always deferred, never arrived at, never finished.

Dramatic heterotopia

Scholars have focused on the heterotopic character of events, and notably festivals as sites facilitating transformations in time, space, and culture. Intending to shape a coherent framework for festival analysis, Quinn and Wilks (2017) configure principles of "festival heterotopia." Given that transformation is a possibility native to heterotopia – inchoate and cacophonous spaces that are potent liminal sites – this effort is welcome. The study of evental heterotopias remains fragmentary, however, with events possessing intentionally transformative agendas notably understudied. The investigation of performative repertoires serves to clarify the complex architectonics of "transformational" events.[12]

As a vast stage for dramatising paradox through performance art, BRC offers a unique case study. The event-city has evolved as a site of dialogue between ludic/civic, participant/spectator, and artist/tourist elements. Although no other event may have performed paradox on this scale, the tendency is apparent in experimental gatherings with transformational agendas. While the concept of "alternative cultural heterotopia" (St John 2001) enabled comprehension of alternative lifestyle gatherings as sites of contested discourse and practice, the concept of "dramatic heterotopia" aids conceptualisation of an event-community cognisant of the role of the arts in redressing its own conflicts. While scholars of Burning Man have addressed the significance of performance in an event where collaborative and interactive repertoires are redolent (see Clupper 2007; Bowditch 2010), the present study is concerned with BRC's unique status as participatory spectacle.

The popularisation of Burning Man has ignited a dilemma. BRC strives to be demarcated in a marginal "other" space that is distinct and exclusive, while adopting an inclusive approach, as enshrined in *Radical Inclusivity*, which states that: "Anyone may be a part of Burning Man. We welcome and respect the stranger. No prerequisites exist for participation in our community." That BRC's population is typically 77% white, with above-average tertiary education rates and personal income status, prompts inquiries about the radicality of its inclusivity.[13] But while the playa may be a realm of exclusive inclusivity, it is also a topos of risky aesthetics, a space of "edgeplay," an experimental mode in which authenticity derives from voluntary risk taking and improvisation, notably in the context of experimental theatre and performance (O'Grady 2017). As a space for the "discontinuity and changeability of existence" (Johnson 2013, 794), a site for the dissolution of boundaries, theatre, and notably participatory theatre provokes understanding of heterotopia as process.

At Burning Man, participants co-create events in a collective approach to event production that ideally converts an *audience* into *performers*. While BRC is a stage for participatory performance, its stature as participatory is fraught with tension. From its desert inception, this unique gathering has been set in a mythically limitless locale for participatory arts. And yet its integrity has been undermined by the spectre of the "spectacle," by the circumstance in which art grows removed from its audience and a divide opens between classes of eventgoers characterised by their status as artists, producers, or volunteers on the one hand, and consumers, clients, and customers on the other. This divide between eventmakers and eventgoers belies the tension at the core of an event reliant on volunteerism *and* commerce.

We have seen that this tension is foundational to Burning Man. In the mid-1990s, the event appeared to have devolved "into yet another commodified, fetishised spectacle of late capitalist culture, to be consumed like a professional sporting event or some kind of desert Lollapalooza" (Wray 1995). While Harvey has poached from Debord's provocations in *Society of the Spectacle*, the BMP is not anti-capitalist and BRC hardly constitutes a proletarian insurrection. Individual participants and theme camps may subscribe to radically autonomous agendas, but resistance to "the spectacle" is foremost theatrical, not ideological. When a banner reading "No Spectators" was raised at Center Camp in the early 1990s, it was not a Marxist slogan but a call for the inclusive creation (and destruction) of art; a call to break the "fourth wall" not smash the bourgeoisie. Under the direct influence of the Cacophony Society, whose motto was "you may already be a member," participants were free to make a spectacle of themselves. The irony of BRC is that it is "a spectacle *par excellence* . . . a spectacle that requires our active involvement for it to exist at all" (Wray 1995). This echoes the stance of Harvey, who wanted to think of the playa

> as a sort of movie screen upon which every citizen of Black Rock City is encouraged to project some aspect of their inner selves. This novel use of nothingness elicits a superabundant production of spectacle. But it is spectacle with a difference. We have, in fact, reversed the process of spectation by inviting every citizen to create a vision and contribute it to a public environment.
>
> (Harvey 2000)

Redressive artopia

If the playa is an event-space bristling with paradox, this appears most evident in its fragile status as a participatory spectacle. While the spectre of the spectator continues to haunt Burning Man, the artifice of BRC enables its inhabitants to navigate its implicit contraries, a process whereby they're acculturated to a principled ecology. Turner's processual "social drama" model serves to illustrate how burner culture is reaffirmed through attention to three art projects that have dramatised event contradictions. Turner demonstrated that social drama – with four phases of breach, crisis, redress, and resolution (reintegration or schism) – is universal to social life. *Breach*: conflict is triggered by a breach of rules, norms, or principles. *Crisis*: escalating controversy involves the wider community. *Redress*: building conflict provokes redressive mechanisms, "ranging from informal arbitration to elaborate rituals, that result either in healing the breach or public recognition of its irremediable character" (Turner 1985, 74). *Resolution*: either in the form of reintegration or schism (e.g. splinter groups). From traditional African tribe to modern community, for Turner (1982), societies evolve performative frameworks – from ritual to theatre – that enable a community to negotiate social tensions – dramas that reaffirm ultimate concerns and restore values at the heart of the human enterprise. Turner focused on the redressive potential of cultural performances ("cultural drama") – how art imitates life, already infused with art. BRC offers a unique exemplar of this perennial cycle. As a "city" built on the foundations of the arts across myriad media and performance frameworks, Burning Man (BRC and its regional progeny) is an evolved site of reflexive potential. That is, it has honed art forms from fire-arts to satire that enable the community to contemplate core values and principles in the face of threats and adversities.

Tourists are iniquitous figures in bohemian, subcultural, and countercultural scenes, often considered parasitical *others* to genuine participants. Typically, scene natives must establish and consolidate their status, and social capital, as authentic insiders by actively identifying this maligned alien. In burner culture, evoking detached spectatorship, the "tourist" is a focal point of self-reflexive satire, often lambasted in BRC publications like *Piss Clear* and *BRC Weekly* and in discussion threads on the typically snarky *ePlaya*. Here, the figure may be excoriated as "sparkle ponies," "broners," or simply "douches." In folk condemnations, the tourist holds parity with the *flâneur* who strolls the space of an exhibition, who is a consumer of trending entertainments, a client to service professionals in the experience industry. As the embodiment of the spectator, the invasive tourist has been identified in a variety of guises, notably the "raver" whose participation has been called into question (see St John 2017a). With increasingly self-indulgent "bucket listers" occupying the playa, and with the self-entitled actions of a growing population showing scant regard for principles like *Civic Engagement, Leaving No Trace,* and *Decommodification,* the limits of *Radical Inclusion* have been tested.

Tensions peaked in 2014 with the "sherpagate" controversy, named after the "sherpas" reported in the *New York Times* (Bilton 2014) to be among the service industry professionals servicing wealthy burners. A flurry of reports ensued on the means by which principles like *Radical Self-reliance* and *Gifting* were being outsourced to the operators of "turnkey" (or "Plug 'n' Play") camps, where clients luxuriate in gated RV compounds with high-end concierge services that appeared to be sanctioned by the BMP. Among those targeted was James Tananbaum, billionaire founder and CEO of leading healthcare investment fund Foresite Capital, and member of the BMP Board of Directors. Tananbaum came under increasing pressure and resigned his position on the Board after his 2014 theme camp was exposed as an enclave for wrist-banded VIPs who arrived on private jets, paid $15K per head, and were to be issued popsicles to be distributed as "gifts" (Gillette 2015). Exposure of prepackaged camps appeared to signal new levels of stratification, with affluent members of the Silicon Valley tech industry and a slew of celebrity attendees at one extreme and their service providers at another. That the ticketing structure permitted operators to purchase blocks of tickets on behalf of their clients suggested a degree of complicity on the part of the BMP. Constituting a "breach," this exposé sparked outrage across social media and fuelled debate in the bloggoverse, triggering fresh lamentation on the demise of Burning Man. A controversy brewing for years erupted into a "crisis" expressed in public grievances, resentment, and recriminations over opportunistic vender-client relationships and the apparent outsourcing of event principles.

The perceived threat to the community demanded a response, with the BMP interrogating intentions, reforming theme camp registration criteria, placement policy, and ticketing sales structure, while limiting reliance on outside service providers and even banning offending and noncompliant theme camps (Goodell 2019). While organisational responses are integral to a redressive climate, my focus is collaborative and interactive art projects. The three projects selected for discussion are the 2015 art theme Carnival of Mirrors, the "mutant vehicle" The Bleachers, and the large-scale installation No-One's Ark (at Israel's Midburn).

Since The Inferno of 1996, art themes have been platforms for expressing the anxieties, hopes, and dreams of the Burning Man community. Widely interpreted, themes can illustrate how Burning Man events are also redressive contexts for the performance of wider societal dramas.[14] In 2015, Carnival of Mirrors offered a self-reflexive tribute to the legacy of the participatory spectacle, where the spectator can become a co-producer, the tourist a creator, the virgin a burner. Given its allusion to de-masking and authentic expression, it was among the event's most self-conscious themes. The narrative epilogue authored by Larry Harvey and Stuart Mangrum sought to distinguish the classic sideshow carnival – where "carnies" are distinguished from members of the public considered "chumps and suckers, marks and rubes" – from Burning Man. "The wall dividing the observer from observed will disappear, as by an act of magic; through the alchemy of interaction, everyone at once can be the carny and the fool." In the wake of ticket scarcity and an insurgence of virgins, Carnival of Mirrors behaved like a billboard for *Radical Inclusion*. At the same time, it appeared to offer a rejoinder to a complaint escalating in the wake of "sherpagate," that BRC had devolved into an entertaining spectacle. Debord supplied the epigraph: "The spectacle is not a collection of images; it is a social relation that is mediated by images." The comment offered a statement on the society that Harvey and others had long imagined Burning Man to be departing from – a society where images, signs, and commodities mediate social relationships. While critics have panned BRC for becoming another Lollapalooza or Coachella, with the Man derogated as a symbol of separation and the main attraction in a new American theme park, under the logic of the Carnival of Mirrors, the effigy appeared to serve as a mirror to eventgoers, likened to participants on a vision quest. The Man pavilion and perimeter was designed such that – having negotiated the bustling Mystic Midway with its booths, "barkers," and "freak shows" and navigated the disorienting maze of the Funhouse – seekers approached a courtyard where they could stand before the towering effigy. There, "the brittle mirror and the occulting mask will melt away, and at this point there'll be no gag, no swag, no souvenir of self; the show will be you."[15] This narrative raised the curtain on the event's inclusive logic: *you are the spectacle*. "What we do is antithetical to the spirit of consumerism," Harvey has stated. "We are the absolute obverse of a theme park. In our theme park, you are the theme" (in Beale 1996).

A response to the possibility of BRC's devolution into a domesticated music festival, or indeed a professional sporting event, The Bleachers (2013–2016) provided a smaller scale entry point to participatory spectatorship. The Bleachers was a "mutant vehicle," or art-car, co-created by a crew based in Vancouver and NYC to resemble a set of sports bleachers. Built on a flatbed truck complete with a tall stand of benches and a scoreboard, the mobile stadium transported burners across the playa and through the backstreets of BRC. Spectating participants were exposed to the commentary of comic announcers occupying an elevated booth and behaving like sportscasters and radio DJs. Operating twenty-four hours over seven days for four consecutive events, among the fifty volunteers were those acting as improv referees, cheerleaders, and ringmasters who whipped the stand into a cheering mass. Occupants were treated to satire drawing attention to absurdities in the playascape. As a form of counter-spectacle in a desert of the surreal, The Bleachers was rich in irony – notably that on-board spectators were also performers on a mobile stage. Such was the intent of project creator and improv artist Neil Pegram, who had long meditated upon the ideal vehicle to stage interactive commentary on the "spectator." Pegram was driven by a few key questions. If burners are sitting on a set of sports bleachers and watching Burning Man, "are they participating or are they spectators?" And "what is the show?"[16] As co-creator Eric Holt adds, the project made people "think a little bit about the paradox of Spectating and Creating, and the interplay between the two" (Gotham 2015). The wood components of the vehicle were ritually burned on-playa in 2016, with Pegram commenting that the project was terminated partly due to people failing to understand the participatory nature of an event reliant on volunteerism and gifting. Among the most inclusive and interactive art cars in the history of BRC, The Bleachers was swamped in its final years by "virgins" who "expected to be driven around and entertained. . . . Kind of like if you went to Coachella and they had a shuttle bus tour." This conclusion points to unresolved concerns impacting the community in the wake of "sherpagate."

No-One's Ark was built and destroyed in 2016 at Israel's regional Burning Man event Midburn, in the Negev Desert. Situated at the margins of the Midburn City "playa," No-One's Ark was a large-scale replica ark funded from multiple sources and constructed from timber by a team of collaborators. The climbable installation was thirty meters long and eight meters high and featured an interior experimental sound environment. As the project's title suggests, No-One's Ark celebrated, and subverted, Jewish heritage (possessed by "no-one," it belonged to everyone). The project was the brainchild of civil engineer Alon Halamit, who had first attended Midburn in 2015 as a "virgin." Expressing his status as a virtual tourist: "I came to consume, I didn't come with anything. I didn't even have a tent, I came with half bottle of rum." Of his first event, Halamit remarked that it "opened my eyes to a different world, a different way of thinking."[17] He claimed to be truly "enlightened" and compelled to give back to the community. Torched in a conflagration on the final night, No-One's Ark was an exemplary gift as it was "property that perishes." The words are those of Lewis Hyde (1983, 10) for whom a true gift is consumed in practices that ensure its spirit remains in circulation (St John 2017b). Destroying one's labours by fire ensures that an object is not possessed, but potentially sparks gratitude among those who bear witness. This sensibility is expressed by Halamit:

> I don't see it as a burn, I see it as an ignition. . . . When something burns inside of you, you can do anything. We wanted to pass this ignition with the burn of the ark to all the people that came to this event saying "People let yourself burn. Let your internal fire Burn. Believe in yourself. Dream. Believe you can fulfill your dreams."

The labours of Halamit and his build team are those of persons afflicted with gratitude. Within a movement climate in which principles like *Gifting* have been undermined due to a growth in convenience culture, this was a project reaffirming core values.

As these projects illustrate, Burning Man is an inclusive culture in which art projects are pivotal to transforming outsiders into insiders and spectators into participants, and restoring the value of beleaguered principles. They emerge within an event-culture that has evolved a repertoire of collaborative and interactive practices calling attention to its own besieged values. The 2015 art theme Carnival of Mirrors intentionally traversed the borders distinguishing spectators from participants, revealing a founding paradox. The Bleachers literally mobilised the potential for eventgoers to perform this paradox. No-One's Ark illustrated how the project's lead designer and his team had been converted from virtual tourists into burners, from consumers into makers. Materialising within an event-culture that encourages individuals to express their authentic selves in unconditional actions, these examples are among many that illustrate Burning Man's stature as a participatory spectacle.

Consistent with the unfinished qualities of radical openness, this drama is still unfolding. The entire processual mechanism Turner identified is evident, with the "crisis" that escalated in the wake of the "sherpagate" "breach" prompting "redressive" practices, including the initiatives discussed. Such projects have been integral to the perpetuity of Burning Man. In Turner's terms, the projects are integral to the "reintegration" of cultural principles – both in BRC and in a multitude of regional events – that demonstrate multiple forms of "redress." An adequate assessment of the career of this social drama, including its "resolution" phase, requires detailed exposition of the Burning Man movement – the intent of a comparative ethnography of burner culture in BRC, Europe, and Israel.[18] Given that Burning Man involves a global network of 90+ affiliated events – a global archipelago of heterotopias – that mimic, mutate, and resist the "mother" event, the project aims to provide further insights on the heuristic value of heterotopia. As regional events are "effectively realized utopias" in which BRC is "simultaneously represented, contested and inverted" (Foucault 2008, 16), like a hall of magic mirrors that replicate, distort, and subvert the prototype, they offer a profusion of solutions to the concerns condensed in this and other crises.

Conclusion

Black Rock City is a distinct context for the performance of paradox endogenous to other spaces and spaces of otherness: a "dramatic heterotopia." While total comprehension remains elusive, familiarisation with this participatory event-culture requires exposure to a complex ethos where principles exist in a fraught relationship. As promulgated by Larry Harvey, the playa is a space infused with possibility arising from paradoxes inhering in an enacted philosophy. While Burning Man is a contested event-space invested with multiple meanings, it features co-creative performance repertoires serving to dissolve difference, resolve tension, and acculturate participants to event principles. The "sherpagate" drama illustrates that Burning Man is a ripe context for the embodiment of contraries. That the playa is pregnant with paradox that can be enacted is pivotal to the transformative potential of this prototypically "transformational" event.

Evental heterotopias are as much processual as spatial. Sherpagate represents a condensation of concerns besetting the burner community. Animated by trenchant discord contesting the significance of Burning Man, the controversy is interwoven with surrealist and civilising tendencies. Notably in its redressive phase, this artopian social drama conveys the performativity of a participatory arts event where binaries like self/other, consumer/producer, tourist/artist, and spectator/participant dissolve. Negotiating this terrain, the art projects discussed served to enable transformative potential while reaffirming undermined principles. This research on the dramatisation of paradox and performance of ambivalence in the city-like "other space" of a dramatic heterotopia furthers understanding of the logics of transformation in such events.

Notes

1 The chapter derives from qualitative research (including field interviews and participant observation) in eight visits to Black Rock City since 2003 and in research conducted in my capacity as Senior Research Fellow for the Swiss National Science Foundation funded project Burning Progeny: The European Efflorescence of Burning Man (January 2016–December 2019) based in Social Science, University of Fribourg, Switzerland: www.burningprogeny.org.

2 The Ten Principles are *Radical Inclusion, Gifting, Decommodification, Radical Self-reliance, Radical Self-expression, Communal Effort, Civic Responsibility, Leaving No Trace, Participation* and *Immediacy* (for an explanation of each, see: https://burning man.org/culture/philosophical-center/10-principles). (Accessed January 15, 2019). In this chapter, all principles are capitalised and italicised.

3 Dehaene and De Cauter's comment in a note on their translation of Foucault's "Of Other Spaces" (Foucault 2008, 25, n15).

4 According to the Black Rock City Census 2013–2017 Summary Report: https://drive. google.com/file/d/1pXuvtM065ZYDGBVzo-mTcXgTotGTnbEX/view.

5 Black Rock City Census: https://burningman.org/event/volunteering/teams/census/.

6 The idea of Burning Man Education Director, Stuart Mangrum, The Burning Man Phrase Generator was originally featured on the Burning Man website, and was published in *The Black Rock Gazette* (September 5, 1999, 2).

7 In this chapter, Black Rock City is used interchangeably with "playa."

8 As listed in the "Out / In" section of *BRC Weekly* 2018.

9 As announced in the 2018 Survival Guide: https://survival.burningman.org/leave-no-trace/.

10 The passage is from Dehaene and De Cauter's clarifying comments to their translation of Foucault's original text.

11 Burning Man Philosophical Center: www.burningmanproject.org/programs#philosophical.

12 For an elaboration of the usefulness of heterotopology for festive studies, especially "transformational" events, see St John (2020), which navigates Burning Man by way of Foucault's six "principles" of heterotopia.

13 Population data from the 2017 Black Rock Census indicated a population including 77% white, 5.6% Asian, 4.9% Hispanic/Latino, 1% black, and 9.3% "other or multiple."

14 A Burning Man-inspired event driven by overt political themes, Catharsis on the Mall, is noteworthy in this regard (see St John 2019b).

15 2015 Art Theme: Carnival of Mirrors. http://burningman.org/culture/history/brc-history/ event-archives/2015-event-archive/theme/.

16 Neil Pegram, communication with the author, April 1, 2019.

17 Alon Halamit, interviewed by the author on Skype, November 16, 2016.

18 See note 1.

Works cited

Beale, Scott. 1996. "Larry Harvey Interview at SOMAR (July 1996)." Published on You-Tube, July 13, 2018. www.youtube.com/watch?v=BZwUC1HHwm8.

Bey, Hakim. 1991. *TAZ: The Temporary Autonomous Zone – Ontological Anarchy and Poetic Terrorism.* New York: Autonomedia.

Bhabha, Homi, K. 1994. *The Location of Culture.* Hove: Psychology Press.

Bilton, Nick. 2014. "A Line Is Drawn in the Desert: At Burning Man, the Tech Elite One-Up One Another." *New York Times,* August 20. Accessed April 29, 2019. www.nytimes.com/2014/08/21/fashion/at-burning-man-the-tech-elite-one-up-one-another.html?ref=fashion&_r=2.

Bonin, Olivier, dir. 2009. *Dust & Illusions.* Madnomad Films.

Bowditch, Rachel. 2010. *On the Edge of Utopia: Performance and Ritual at Burning Man.* Chicago, IL: University of Chicago Press.

Clupper, Wendy. 2007. "The Performance Culture of Burning Man." PhD diss., Department of Theatre and Performance Studies, University of Maryland.

Davis, Erik. 2005. "Beyond Belief: The Cults of Burning Man." In *Afterburn: Reflections on Burning Man,* edited by Lee Gilmore and Mark Van Proyen, 15–40. Albuquerque, NM: UNM Press.

Debord, Guy. 1970 [1967]. *Society of the Spectacle.* Detroit: Black and Red.

De Cauter, Lieven, and Michiel Dehaene. 2008. "The Space of Play: Towards a General Theory of Heterotopia." In *Heterotopia and the City: Public Space in a Postcivil Society,* edited by Michiel Dehaene and Lieven De Cauter, 87–102. Abingdon: Routledge.

Dehaene, Michiel, and Lieven De Cauter, eds. 2008. *Heterotopia and the City: Public Space in a Postcivil Society.* Abingdon: Routledge.

Doherty Brian. 2006. *This Is Burning Man: The Rise of a New American Underground.* Dallas, TX: BenBella Books.

Evans, Kevin, Carrie Galbraith, and John Law. 2013. *Tales of the San Francisco Cacophony Society.* San Francisco: Last Gasp Publishing.

Falassi, Alessandro. 1987. "Festival Definition and Morphology." In *Time Out of Time: Essays on the Festival,* edited by Alessandro Falassi, 1–12. Albuquerque, NM: UNM Press.

Foucault, Michel. 2008 [1967]. "Of Other Spaces." Translated by Lieven De Cauter and Michiel Dehaene. In *Heterotopia and the City: Public Space in a Postcivil Society,* edited by Michiel Dehaene and Lieven De Cauter, 13–29. Abingdon: Routledge.

Fox, William L. 2002. *Playa Works: The Myth of the Big Empty.* Reno, NV: UNP.

Garrett, Rod. 2010. "Designing Black Rock City." *Burning Man Journal,* April 20. Accessed July 20, 2017. http://journal.burningman.org/2010/04/black-rock-city/building-brc/designing-black-rock-city/.

Gillette, Felix. 2015. "The Billionaires at Burning Man." *Bloomberg Businessweek,* February 5. Accessed July 7, 2017. www.bloomberg.com/news/articles/2015-02-05/occupy-burning-man-class-warfare-comes-to-desert-festival.

Gilmore, Lee. 2010. *Theater in a Crowded Fire: Ritual and Spirituality at the Burning Man Festival.* Berkeley: UC Press.

Gómez, Manuel A. 2013. "Order in the Desert: Law Abiding Behavior at Burning Man. *Journal of Dispute Resolution* (2): 349–73.

Goodell, Marian. 2019. "Cultural Course Correcting: Black Rock City 2019." *Burning Man Journal,* February 9. Accessed April 29, 2019. https://journal.burningman.org/2019/02/philosophical-center/tenprinciples/cultural-course-correcting/.

Gotham, Terry. 2015. "Why We Burn – Eric" (interview with Eric Holt). *Burners.me*, November 26. Accessed April 29, 2019. https://burners.me/tag/the-bleachers/.

Harvey, Larry. 2013. "Introduction: The Philosophical Center." *Burning Man Journal*, November 12. Accessed April 29, 2019. https://blog.burningman.com/2013/11/tenprinciples/introduction-the-philosophical-center.

———. 2000. *La Vie Boheme: A History of Burning Man*. Minneapolis, MN: Lecture at the Walker Art Center, February 24. Accessed April 29, 2019. https://burningman.org/culture/philosophical-center/founders-voices/larry-harveys-writings/la-vie-boheme/.

Hetherington, Kevin. 1997. *The Badlands of Modernity: Heterotopia and Social Ordering*. London: Routledge.

Hyde, Lewis. 1983. *The Gift: Imagination and the Erotic Life of Property*. New York: Vintage.

Johnson, Peter. 2013. "The Geographies of Heterotopia." *Geography Compass* 7 (11): 790–803.

Maffesoli, Michel. 1996 [1988]. *The Time of the Tribes: The Decline of Individualism in Mass Society*. London: Sage.

Massey, Doreen. 2005. *For Space*. London: Sage.

Mitchell, Jon. 2018. "Mayors of U.S. Cities Visit Black Rock City." *Burning Man Journal*, October 23. Accessed April 29, 2019. https://journal.burningman.org/2018/10/news/brc-news/mayors-of-u-s-cities-visit-black-rock-city/.

O'Grady. Alice. 2017. "Introduction: Risky Aesthetics, Critical Vulnerabilities, and Edgeplay: Tactical Performances of the Unknown." In *Risk, Participation, and Performance Practice: Critical Vulnerabilities in a Precarious World*, edited by Alice O'Grady, 1–29. Cham, Switzerland: Palgrave Macmillan.

"Out/In." 2018. *BRC Weekly*, August 27–September 2, issue 9, 2. Accessed April 29, 2019. https://brcweekly.com/BRCWeekly2018_int.pdf.

Pike, Sarah M. 2011. "Burning Down the Temple: Religion and Irony in Black Rock City." *Religion Dispatches*, September 11, 2011. Accessed April 29, 2019. www.religiondispatches.org/archive/culture/5082/burning_down_the_temple%3A_religion_and_irony_in_black_rock_city.

———. 2001. "Desert Goddesses and Apocalyptic Art: Making Sacred Space at the Burning Man Festival." In *God in the Details: American Religion in Popular Culture*, edited by Eric Mazur and Kate McCarthy, 155–76. New York: Routledge.

———. 2000. "The Burning Man Festival: Pre-Apocalypse Party or Postmodern Kingdom of God?" *The Pomegranate: A New Journal of Neopagan Thought* 14: 26–37.

Quinn, Bernadette, and Linda L. Wilks. 2017. "Festival Heterotopias: Spatial and Temporal Transformations in Two Small-Scale Settlements." *Journal of Rural Studies* 53: 35–44.

Rohrmeier, Kerry, and Paul Starrs. 2014. "The Paradoxical Black Rock City: All Cities Are Mad," *Geographical Review* 104 (2): 153–73.

Saunders, Teddy, Parker Howell, and William Walsh, dirs. 2012. *Oh, the Places You'll Go at Burning Man*. Tedshots Productions, Inc. www.youtube.com/watch?v=ahv_1IS7SiE.

Scaruffi, Piero. "John Law Interview: Part3 Burning Man." Published on YouTube November 8, 2015. www.youtube.com/watch?v=NA45PAf7pWc.

Silver, Marc, dir. 2000. *Burning Man: Community or Chaos?* Channel 4.

Soja, Edward. 1995. "Heterotopologies: A Remembrance of Other Spaces in the Citadel-LA." In *Postmodern City Spaces*, edited by Sophie Watson and Katherine Gibson, 13–34. Oxford: Blackwell.

Spariosu, Mihai I. 2015. *Modernism and Exile: Play, Liminality, and the Exilic-Utopian Imagination*. Hampshire: Palgrave Macmillan.

St John, Graham. 2020. "Ephemeropolis: Burning Man, Transformation and Heterotopia." *Journal of Festive Studies* 1(2) Forthcoming.

———. 2019a. "At Home in the Big Empty: Burning Man and the Playa Sublime." *Journal for the Study of Religion, Nature and Culture* 13 (3) Forthcoming.

———. 2019b. "The Cultural Heroes of Do-ocracy: Burning Man, Catharsis on the Mall and Caps of Liberty." *Liminalities: A Journal of Performance Studies* 15 (1). http://liminalities.net/15-1/burning.pdf.

———. 2017a. "Charms War: Dance Camps and Sound Cars at Burning Man." In *Weekend Societies: Electronic Dance Music Festivals and Event-Cultures*, edited by Graham St John, 219–44. New York: Bloomsbury.

———. 2017b. "Blazing Grace: The Gifted Culture of Burning Man." *NANO: New American Notes Online*, 11. Accessed April 29, 2019. https://nanocrit.com/issues/issue11/Blazing-Grace-The-Gifted-Culture-of-Burning-Man.

———. 2001. "Alternative Cultural Heterotopia and the Liminoid Body: Beyond Turner at ConFest." *The Australian Journal of Anthropology* 12 (1): 47–66.

———. 1982. *From Ritual to Theatre: The Human Seriousness of Play*. New York: PAJ Books.

van Gennep, Arnold. 1960. *The Rites of Passage*. Translated by Monika Vizedom and Gabrielle L. Caffee. Chicago, IL: University of Chicago Press.

Wray, Matt. 1995. "Burning Man and the Rituals of Capitalism." *Bad Subjects: Political Education for Everyday Life*: #21.

13 Afterword

Kevin Hetherington

> The sailing vessel is the heterotopia par excellence. In civilisations without ships the dreams dry up, espionage takes the place of adventure, and the police that of the corsairs.
>
> Michel Foucault, "Different Spaces" (1998, 185)

As I write this afterword in the end of summer days in August 2019, I ponder on two seemingly unconnected things: the purpose of an afterword in the context of a book on heterotopia and what is happening out in the North Atlantic this week.

The historian of paratexts, which is a term for the bits of a book like the preface, postface, index, title, acknowledgements, etc., outside the main body of the text, Gérard Genette (1997), tells us that afterword or postface are rare in literature, certainly in comparison with prefaces (which, of course, are usually written afterwards but put before as an introductory guide). Where they do occur it is often as a somewhat ineffective corrective by the author often included to try to stop the reader developing a poor reading that they have perhaps invited through their clumsy writing or poor argument. With an edited book like this it invites the idea of some overall commentary or reflexion, perhaps on a key theme from the book drawn from one or more of the chapters. However, the reader having done all the work to get to that point in the book probably is not really interested in having it re-interpreted for them by the author or someone else now suggesting a different view or adding something else late in the day that was missing from the main body of the text (Genette 1997, 238–39). To do so suggests that the reader, assuming they have followed the logical order in their reading of the text, should go back and start again or at least re-read parts of it. Likewise, an afterword does not really fulfil another possible function of staving off an anticipated hostile reception by the critics by getting the author's defence in first. The critics are still likely to come to their own view on the text despite that author trying to give a very visible, (defensive?) late steer. As Genette (1997, 239) says, "But for the postface, it is always too early and too late." The issue is, then, how to steer something that either does not want to be steered or is rather difficult to do so?

I am not, though, the author of this book – just another contributor. However, I am one of the authors who, over the last couple of decades, has helped establish a reading of Foucault on heterotopia (Hetherington 1997, 2011, 2015), so perhaps this afterword is as much to my own work on the subject as it is to this particular volume. After all, an invite to make such a contribution usually relates to some issue of intertextuality – having something of a relevant reputation to speak on a topic with the possibility that I may have something to add. One speaks from author-ity.

However, there is something more pertinent to the issue in question at work as well. Although Genette does not say this himself it is clear that paratexts – and not just afterword and prefaces but all the supplementary texts in a book that are outside the main body of the text – index, title page, acknowledgements, dedications, forewords and so on – can have a heterotopic relationship to the content of the book in which they are bound. They are not the book itself but add-ons whose emplacement relates to but also does not relate to the core work in that it stands somewhat outside of the interpretation that is otherwise given in the body of the chapters. And that relationship can be established by author or audience in their writing/reading of what is placed outside of the main text but in relation to it. It suggests that other readings are possible, that there is always a supplement and that things remain unfinished despite the craft of the author/editors.

I want to use this opportunity and the space in which it arises to say something about Foucault. I do not intend to give an overall steer on Foucault's sketchy essay on heterotopia nor on its now substantial secondary commentary, including in this volume which tries to situate the term in relation to a troubling time in the history of global relations and their expressions of power and control. There is one theme, though, I do want to draw out both in the context of what he said there and the position of that text within his work as a whole (see also Deleuze 1988; Shapiro 2003; Hetherington 2011).

Foucault's essay on heterotopia deals with both places, real and imagined sites or types of site, and the idea of emplacement that is relational (1998). The essay, where he tries to define what he means by heterotopia, mingles together, then, a description of particular types of outsider place: cemeteries, museums, brothels, gardens and so on, with an idea of space as relational through the notion of emplacement established through a set of six principles. Overall, and in a somewhat chaotic and not fully developed way, what he is trying to achieve is an understanding of the modern and its spatial dynamic (for him that means European modern) by uncovering a logic of emplacement within these places and the mirror that they hold up to society as a whole as it changes. He chooses sites that are somehow other to the recognised or familiar discourse of spatiality of their time.

The confusion in his essay, and the fact that no one has ever subsequently fully disentangled it or come to a definitive reading accepted by the many, is that he does not really say why these spaces and what relationship they establish to the idea of a wider sense of society. The closest he gets is through using the metaphor of a mirror, in the sense that these are spaces that are other and reveal something about what is understood as same through the relationship they establish. Same/other, normal/pathological, madness/reason – much of Foucault's early work establishes something of this reflective dynamic between these binarisms.

Overall, as Deleuze (1988) was first to really point out, what he is exploring is the relationship between the discursive and the figural or between saying and seeing and their different modes of operation within knowledge (see also Shapiro 2003).

As Shapiro (2003) has pointed out, Foucault's work can be seen as developing through three distinct periods and objects of enquiry. The first is his "structuralist" phase with his interest in how discourse is established from speech, the main example being his work *The Order of Things* (1989b [1966]) which focuses primarily on textual sources found in the archive. The third phase is a phase where he becomes interested in the visibilities of power and with mode of governing in societal spatial locations and not just in the archive. The major statement here, of course, is his work *Discipline and Punish* (1977 [1975]) which is concerned with the physical spatial apparatuses of panoptical power within carceral institutions like the prison. The shift between these two phases is often marked out as a shift of emphasis either from archaeology to genealogy (Dreyfus and Rabinow 1982) or from the archive to the map (Deleuze 1988).

But what of the time in-between, the second phase from 1967–74 or thereabouts? This, of course, is the period in which the heterotopia essay was written. What we see in this period is Foucault moving away from the archival sources but not yet moving on to look at the carceral institutions as spatial apparatuses. The main work of this period is the now largely forgotten/unfashionable *Archaeology of Knowledge* (1974 [1969]). We know that in this time Foucault also planned a book on Manet and modern art but only one lecture survives (2009). We know it was also a time of growing political activism for him after the events of 1968 in Paris and his absence from them and his reappraisal of the mechanisms of power within modern society.

While a rather dry and abstract book, *The Archaeology of Knowledge* does introduce a range of interesting concepts around the idea of the spatiality of emergence and the operationalisation of knowledge not found elsewhere in which he tries to establish the relationship between what is seen and what is said or between the discursivity of the archive (the known) and the figurality of the map (the as yet unknown but visible) though he was clearly dissatisfied with the outcomes. His ideas on heterotopia are part of this period and are a brief commentary, albeit elliptically, on the relationship between what is seen (places) and what is said (emplacement principles).

His heterotopia essay is about where things go and what stands out from that and the complexity that surrounds this issue in the present time. Many emplacements are knowable and understandable – they exist within a discourse that is familiar but others are not, they relate figurally as knowledge by being seen rather than discursively as knowledge – understood within a recognisable system of emplacement. Some of these spaces, he suggests, are imaginary – utopia – and others are real sites – those he calls heterotopia (1998, 178). He then goes on to offer a sketch outline of the study of such spaces, or what he calls an heterotopology (1998, 179), giving brief examples and the effects they can have in allowing us to reflect on wider issues of spatiality within society and the social relations that inform it.

This is one of his first attempts to map out issues of spatial relationality. He had tried to do something similar in the first chapter of *Madness and Civilisation*, writing on the leper colony and the ship of fools (1989a [1961]) and using there the anthropological language of liminality. In his heterotopia essay, though, he focuses on identifying the principles of heterotopia of which he notes six: spaces of deviation, spaces with changing function, spaces of juxtaposition of the incongruous, spaces of temporal discontinuity (anachronistic), spaces that are both open/closed to entry, and spaces of contrast or difference (1998). Many of the chapters in this present book light on one or more of these for their analysis through a particular example, real or literary/artistic. What Foucault does not say is why these principles and what their societal functions are – he has not yet developed the theoretical vocabulary for his heterotopology, and he takes it no further in his work.

Nor does he carry out what would later come to be called a genealogy of the sites he associates with these principles. The suggestion at this point is that it might be a direction he will take – the library and gallery, in particular, were spaces that interested him at this time – though clearly he does not. What he does do, as Deleuze cogently puts it, is go on through later work to establish a cartographic understanding of the diagram or apparatus of power (1988) as inherently spatialised and that then becomes his focus in his later work on the panoptical carceral institutions of power such as the prison and what they have to tell us about power more generally and its role in constituting modern subjects (1977). In many respects his essay on heterotopia is an early example of such an approach. For some, its unclear formulation or attempt to discursively name figural otherness and fix it are problematic (see Genocchio 1995; Saldanha 2008). For others, including those in this volume, the lack of a systematic theoretical toolkit leaves open a range of possibilities for its use.

While the language and the mode of analysis may be incomplete in this period of his work, it does retain a playfulness and a space for play that is lost in the later carceral work of mapping out the apparatus and materiality of power within modernity. We find that more clearly when he speaks of surface of emergence in *The Archaeology of Knowledge* (1974; see also Elden 2001). Without going into a lengthy discussion of that work (one of the freedoms of an afterword, dear reader, is that I do not have to) in a nutshell a surface of emergence is defined as a space in which the already visible but not yet known becomes knowable discursively for the first time – or, where the figural becomes discursive (1974, 41ff). And that happens in a diverse set of spaces, though his own interest has always circulated around the asylum, clinic and prison.

While I am not saying that heterotopia and surface of emergence are the same concept – they clearly are not – they do have a kinship; they perform a very similar function for Foucault in opening up to scrutiny in the real world the relationship between what is seen and how it becomes known and the effects that that knowledge can then have. That, too, remains a theme of many of the chapters in this present book. The prosaic and colourful examples associated with heterotopia are lost in this more abstract later term but whichever term we use, both explore the principle of placement that is formed out of the dynamic between what is seen and what is said in a way that leaves open different possibilities to the institutionalised world he was to go on to describe a few years later.

If an afterword can be said to try and give a steer before/after the main reading, can the same be said for heterotopia as a broader idea? I think so. It can hold a mirror up to society, but it also can show a glimpse of alternate possibilities. It is not all nailed down, or laid out. It does not provide a clear message, rather a picture of one. Like a ship we can tack. That fluidity, like the meandering sailing vessel, is a virtue rather than a flaw.

Steering, then, is also something as a metaphor very much tied up with the imagery of the boat – a theme that fascinated Foucault from his early work on madness and his discussion of the Renaissance ship of fools (1989a, 7–19) through to, and including, the rather cryptic passage that comes at the very end of his heterotopia article (1998) that is the starting point for the chapters in this volume. It is perhaps fitting that he ends his rather elliptical essay on heterotopia (first given as a lecture to some architects) with the idea of the ship. He says no more about it in his heterotopia essay, but it is a major trope in *Madness and Civilisation.*

Whether the ship of fools in the fifteenth and sixteenth centuries were historically real or not (Foucault is a little vague on his sources on this) they had a real cultural presence in the writing from a book of that title by Sebastian Brandt from 1494 and later in the etchings of Albrecht Durer and paintings of Hieronymus Bosch. All, no doubt, were aware, too, of the unruly ship in Plato's *Republic* as a source of reference in drawing up this space of contrast that supposedly sailed up and down the great rivers of central Europe full of the unruly and mad – making visible what could not yet be spoken discursively.

This carnivalesque metaphor, challenging Catholic orthodoxy, not only came into being after the closure of the leper colonies, as Foucault points out, but only 50 years after the beginnings of printing and at a time when the first pre-discursive, pre-Lutheran protestant stirrings must have been in the air. Foucault speaks of this ship as new form within the imaginary landscape of the Renaissance, a challenging space of contrast in which madness became visible before it was known as madness.

Centrally Foucault speaks of those on the ship of fools as prisoners of the passage – one whose state is to pass between but never fully arrive (1989a, 11). And that brings me to the North Atlantic – that great puddle of unpacific weather – that has defined the Western spatial imagination since it replaced the Mediterranean of Ancient times during the Renaissance and the era of European globalisation, discovery, conquest and slavery. Two things have happened there this week. The first relates to Greta Thunberg, the 16-year-old Swedish climate change activist, who has captured the imagination of the world since she began her Friday school strikes against the inaction to halt climate change. Over the last year she has exploded in the space of social media and become the focus for a newly emergent climate change protest movement called Extinction Rebellion. She is, as I write, about to complete a two-week journey on a carbon neutral yacht called Malizia 2 (Malice) under the slogan "unite behind the science." She is presently on her way to speak at a conference on climate change in New York and chose this mode of travel rather than an aircraft with all its climate damaging emissions. The second, in complete contrast to Thunberg, is that US President Donald Trump has cancelled a state visit to Denmark after his announcements, also on social media, that he wished to buy Greenland from Denmark were rebuffed. The former Prime Minister of Denmark Lars Lokke Rasmussen spoke for many when he called Trump's suggestion "absurd."

The discourse of globalisation has always been shaped by a westerly spatial imaginary as much as by its attempt at fixing and exoticizing the east as Other to itself in a discourse of orientalism (see Said 1979). The capitalist world has since the 1400s been oriented westward against the prevailing wind both spatially and temporally. When Columbus set sail to look for a western passage he was not seeking to discover the Americas but to find a new route to Jerusalem (see Todorov 1982). The westward search for what lies beyond the horizon is a key trope in the Western discourse of utopia that underlies notions of expansion, development and progress – an imaginary land that might be found just beyond the western horizon (see Marin 1993).

California dreamings: from the frontier and the expansion of the railroads in the nineteenth century to space exploration and the privatisation of the solar system in the dreams of some of today's billionaire tech entrepreneurs, discovering what lies just beyond the horizon – setting sail "westward" in the direction of the setting sun and the forward movement of time, in search of that other space of dreams is what lies at the boundary between reason and madness in the Western capitalist imaginary. It is a space of overcoming, of conquest, following in the wake of the Portuguese Man O'War that first established the charts for this idea of the traversal and the conquest of space (see Law 1986). Madness and civilisation. Always entwined. Both engage with the Atlantic literally and figuratively; "Water and madness have long been linked in the dreams of European [Western] man" (Foucault 1989a, 12).

Thunburg's approach follows this direction of travel while also challenging it – laudable, her rational utopianism is uncompromising in its "folly" of demanding the impossible – total societal change within a decade to meet emission targets that would stop the globe from warming up by more than 1.5C – something that would make capitalism unviable on that kind of time-scale. Her clear voice is very much one grounded in Enlightenment reason: that the scientific evidence and models are telling us very clearly that we have to stop doing what we are doing to our planet if we are to avoid likely species extinction from the consequences of our self-induced climate change. This "sanity" is counterposed to the practices of societal "madness" and is grounded in the scientific rationality of an Enlightenment empiricism around tipping points that are entirely plausible and requiring of us urgent action. Consciousness has been roused. But action beyond the symbolic?

She has chosen her voyage on a sailing vessel that doubles as a ship of science to make this point. Following her on Twitter, to date, her journey seems to have been calm and uneventful, but there will be a storm of a different kind when she arrives – a storm of media attention to amplify the Twitter storm that she has created with her highly effective use of social media within her protest and call for action (see also Johnson, this volume). In total contract in his political values Trump, too, speaks here the social media language of rationality and absurdity but to quite different ends. A man who has made his billions from real estate seeks to do a big real estate deal with another country by buying part of its, albeit autonomous, territory. He wants the natural resources that can be found there and which melting ice and permafrost from climate change will make it easier to exploit. The "folly" of his proposed actions is to challenge the very basis of the political alliance that has held the North Atlantic nations together since 1945, but that seems to escape him. National self-interest, America First, is all that matters and yet the very idea of the USA is grounded within the totality of leadership of that Western world and not its fragmentation. And there is no concern at all for the climate change that has made Greenland an inviting territorial proposition.

Both also speak the language of social media, the 140 character language of Twitter that calls forth bold, forthright, blunt speech acts lacking in subtle discourse. The politics, of course, is very different as are the ethics. Both speak of self-preservation, one on behalf of humanity in the face of its folly, the other of the self-interest of their nation in the face of increasing competition and the threat of declining global influence. And yet both emerge from a state of crisis and of the changing consequences of globalisation. The environmental impacts of folly capitalism – and let's call it that rather than invoking the false consciousness of speaking instead of the folly of humanity or of "this generation" – and the crisis in the Western alliance that is emerging out of the economic crisis of 2008/9 and austerity politics of neo-liberalism which has allowed nationalism and authoritarian populism to return to the political scene are both the space in which Thunberg and Trump exist. Both speak a language of emergency. One, grounded in a discourse of science, that time is running out to do something about retaining the habitability of our planet, the other, a discourse of unfettered capitalism, giving voice to the extremist discourse of the alt-right within the mainstream such that it risks becoming normalised as part of the political discourse for the foreseeable future. Both, then, in their different ways occupy and constitute a space of heterotopia – a space of challenge and contrast to a crisis of globalisation, in particular, just different challenges, different audiences and different desired outcomes. In many respects each is the embodiment of the challenge for the other.

And yet I have to say that as much as I admire Thunberg's challenge, not least to Trump and what he stands for, it is not Thunberg's well-equipped sailing vessel on the Atlantic that is the defining heterotopia of our time of global political, economic and environmental change. Rather it is that set of overloaded small ships on the Mediterranean desperately trying to carry refugees, migrants and the displaced from Africa and the middle-east out of war zones, failed states and areas of drought, famine and resource depletion and into Europe, many of them in the hands of modern-day pirates and slave traders or people smugglers, that take on that role. We see them nightly on television, many of them sinking with hundreds of lives lost each time, some washing up as rotting corpses on the tourist beaches of Europe (see Kluwick and Richter, this volume). We see in them the look of desperation that displacement causes and how that reflects our own actions and responsibilities. And for those who do make it, the modern-day internment camps, detention centres and places for deportation await them as a second heterotopia (Agamben 1998). What is seen and what is said remains disconnected except in a spectacle of horror. And in them we also glimpse a possible future, our future, the future not just of the hundreds of thousands on the move but of countless millions as the truths of global economic crisis, political re-alignment and climate change come to together in their global effects over the course of this century. Here then is the real madness of folly-capitalism and the consequences of environmental change and also the break-up of the Western alliance and its influence through political and social liberalism and values of democracy. It is their visibility that calls forth a need for a new discourse of space no longer defined by the old language of nation-states, territory and boundary but an open space of the Anthropocene in which we seek to accommodate ourselves to a changing and less benign future that we might have imagined only a few years ago.

Trump and Thunberg. Climate change denier and witness. Both use Twitter as a space in which to speak. Both speak very differently. When she lands, I am sure she will have much to say. However, she has said when she arrives in the US she does not intend to speak to him as he would not listen to her. She is right.

I would not try to put words in her mouth, but through her speech and her actions I am reminded of the words of another articulate, teenage girl about another ship that we might want to say to him about his many walls and denials,

> If by your art, my dearest father, you have
> Put the wild waters in this roar, allay them.
> The sky, it seems, would pour down stinking pitch,
> But that the sea, mounting to the welkin's cheek,
> Dashes the fire out. O, I have suffered
> With those that I saw suffer: a brave vessel,
> Who had, no doubt, some noble creature in her,
> Dash'd all to pieces. O, the cry did knock
> Against my very heart. Poor souls, they perish'd.
> Had I been any god of power, I would
> Have sunk the sea within the earth or ere
> It should the good ship so have swallow'd and
> The fraughting souls within her.
>
> *The Tempest*, Act 1 Scene 2

Works cited

Agamben, Giorgio. 1998. *Homo Sacer: Sovereign Power and Bare Life*. Stanford, CA: Stanford University Press.

Deleuze, Gilles. 1988. *Foucault*. London: Athlone Press.

Dreyfus, Hubert, and Paul Rabinow. 1982. *Michel Foucault: Beyond Structuralism and Hermeneutics*. Brighton: Harvester Press.

Elden, Stuart. 2001. *Mapping the Present: Heidegger, Foucault and the Project of a Spatial History*. London: Bloomsbury.

Foucault, Michel. 2009. *Manet and the Object of Painting*. London: Tate Publishing.

———. 1998. "Different Spaces." Translated by Robert Hurley. In *Michel Foucault, Aesthetics, Methods and Epistemology, Essential Works of Foucault, 1954–1984, Volume Two*, edited by James Faubion, 175–98. London: Allen Lane.

———. 1989a. *Madness and Civilisation: A History of Insanity in the Age of Reason*. Translated by Richard Howard. London: Tavistock/Routledge.

———. 1989b. *The Order of Things: An Archaeology of the Human Sciences*. London: Tavistock/Routledge.

———. 1977. *Discipline and Punish: The Birth of the Prison*. Translated by Alan Sheridan. London: Allen Lane.

———. 1974. *The Archaeology of Knowledge: And the Discourse on Language*. Translated by A. M. Sheridan Smith. London: Tavistock.

Genette, Gérard. 1997. *Paratexts: Thresholds of Interpretation*. Translated by Jane E. Lewin. Cambridge: Cambridge University Press.

Genocchio, Benjamin. 1995. "Discourse, Discontinuity, Difference: The Question of 'Other' Spaces." In *Postmodern Cities and Spaces*, edited by Sophie Watson and Katherine Gibson, 35–46. Oxford: Blackwell.

Hetherington, Kevin. 2015. "Foucault and the Museum." In *The International Handbook of Museum Studies*, edited by Andrea Witcombe and Kylie Message, 21–40. Oxford: Blackwell.

———. 2011. "Foucault, the Museum and the Diagram." *The Sociological Review* 59 (3): 457–75.

———. 1997. *Badlands of Modernity: Heterotopia and Social Ordering*. London: Routledge.

Law, John. 1986. "On the Methods of Long Distance Control: Vessels, Navigation, and the Portuguese Route to India." In *Power, Action and Belief*, edited by John Law, 234–63. Henley: Routledge.

Marin, Louis. 1993. "Frontier of Utopia: Past and Present." *Critical Inquiry* 19 (3): 397–420.

Said, Edward. 1979. *Orientalism*. New York: Vintage Books.

Saldanha, Arun. 2008. "Heterotopia and Structuralism." *Environment and Planning A* 40 (9): 2080–96.

Shapiro, Gary. 2003. *Archaeologies of Vision: Foucault and Nietzsche on Seeing and Saying*. Chicago, IL: University of Chicago Press.

Todorov, Tzvetan. 1982. *The Conquest of America: The Question of the Other*. Translated by Richard Howard. New York: Harper and Row.

Index

Note: numbers in *italics* indicate a figure.

Printed in the United States
by Baker & Taylor Publisher Services